STORIED DESERTS

Storied Deserts makes a crucial and critical intervention in the field of environmental humanities by showcasing an emerging body of research on desert places from around the world.

Deserts, despite dominant stereotypes of wasteland and barrenness, are culturally and ecologically abundant places. This edited volume sets out to reimagine the world's desert places and the very concept of "the desert" itself, taking a boldly interdisciplinary and multicultural approach. Authors engage in literary ecocriticism and ecopoetics, film and visual studies, critical theory, personal and transdisciplinary reflection, creative practices, and historical scholarship. Through their diverse range of perspectives, contributors show how arid lands have been and can be understood as sites of narrative production, places where signs and imaginaries are born from the materialities of space and entanglement. In this way, this volume highlights how the storied matter of the Earth's deserts informs lived realities, environmental histories, cinematic and literary imaginaries, political conflicts, and even intellectual categories such as "the human" and "the elemental".

Ultimately, this book shows that reimagining desert places can help us to grapple with the epochal challenges of the Anthropocene. It is an important and engaging collection for scholars and students across disciplines that helps establish the value of desert humanities.

Celina Osuna is a scholar and an artist. She is an Assistant Professor of English at the University of Texas at El Paso, and her research, with an emphasis on Indigenous and Latinx Environmentalisms, explores aesthetics of desert places in literature, art, and film and their impacts on cultural imagination and geopolitical relationships to land. Her monograph, *Desert Distortion,* is forthcoming with Texas Tech University Press.

Aidan Tynan is Senior Lecturer in English literature at Cardiff University. He is the author of two monographs, *Deleuze's Literary Clinic: Criticism and the Politics of Symptoms* (2012) and *The Desert in Modern Literature and Philosophy: Wasteland Aesthetics* (2020). He has published two edited volumes, *Deleuze and the Schizoanalysis of Literature* (2015) and *Credo Credit Crisis: Speculations on Faith and Money* (2017).

Routledge Environmental Humanities

Series editors: Scott Slovic (University of Idaho, USA),
Joni Adamson (Arizona State University, USA) and **Yuki Masami** (Aoyama Gakuin University, Japan)

The *Routledge Environmental Humanities* series is an original and inspiring venture recognising that today's world agricultural and water crises, ocean pollution and resource depletion, global warming from greenhouse gases, urban sprawl, overpopulation, food insecurity and environmental justice are all *crises of culture*.

The reality of understanding and finding adaptive solutions to our present and future environmental challenges has shifted the epicenter of environmental studies away from an exclusively scientific and technological framework to one that depends on the human-focused disciplines and ideas of the humanities and allied social sciences.

We thus welcome book proposals from all humanities and social sciences disciplines for an inclusive and interdisciplinary series. We favour manuscripts aimed at an international readership and written in a lively and accessible style. The readership comprises scholars and students from the humanities and social sciences and thoughtful readers concerned about the human dimensions of environmental change.

For more information about this series, please visit: www.routledge.com/Routledge-Environmental-Humanities/book-series/REH

STORIED DESERTS

Reimagining Global Arid Lands

Edited by
Celina Osuna and Aidan Tynan

Routledge
Taylor & Francis Group
LONDON AND NEW YORK

earthscan
from Routledge

Designed cover image: Getty Images

First published 2024
by Routledge
4 Park Square, Milton Park, Abingdon, Oxon OX14 4RN

and by Routledge
605 Third Avenue, New York, NY 10158

Routledge is an imprint of the Taylor & Francis Group, an informa business

British Library Cataloguing-in-Publication Data
A catalogue record for this book is available from the British Library

ISBN: 978-1-032-50181-9 (hbk)
ISBN: 978-1-032-50179-6 (pbk)
ISBN: 978-1-003-39721-2 (ebk)

DOI: 10.4324/9781003397212

Typeset in Times New Roman
by codeMantra

CONTENTS

CONTRIBUTORS

Andrea Casals-Hill earned her Master's in Human Settlements and Environmental Studies and Doctorate in Literature, both from Pontificia Universidad Católica de Chile (UC), where she currently conducts a three-year research project on YA dystopian narratives funded by the National Agency for Research and Development (ANID); she was also a postdoctoral fellow at UC researching on environmental awareness in Chilean picturebooks (also funded by ANID). She teaches at graduate and undergraduate programs and is coeditor of the academic journal *English Studies in Latin America*, all at Facultad de Letras UC. Casals collaborates ad honorem as an international liaison for the Association for the Study of Literature and Environment (ASLE). In 2019, she published *Futuro esplendor: ecocrítica desde Chile* (Orjikh), coauthored with Pablo Chiuminatto.

Pablo Chiuminatto earned his Doctorate of Philosophy with a mention in Aesthetics and Art Theory and Master of Visual Arts from the University of Chile. He is an associate professor of the Faculty of Letters of the Pontifical Catholic University of Chile and member of the Laboratory of Social Innovation in Visual Disability OCULAB UC. He is also an associate researcher at the Book and Reading Corporation and the Research Network on Environmental Humanities (RIHA-Chile), and editor and author of a dozen titles, the most recent are A. Casals-Hill & P. Chiuminatto, *Futuro Esplendor: ecocrítica desde Chile* (2019); P. Chiuminatto and V. Rosales, *La imaginación: el taller de la mente* (2019); and P. Chiuminatto and Ignacio Veraguas, *Gusto sabor y saber* (2022).

Anna Flores is a border-dwelling poet and writer born in Nogales, Arizona. She holds an MFA in Creative Writing from Arizona State University, where she currently teaches English Composition, Horror Fiction, and Poetry. Her recent awards and honours include a 2021 Swarthout Award in Writing, a 2021 Center for

Imagination in the Borderlands Creative Research Fellowship, and a 2020 Mabelle A. Lyon Poetry Award. You can read her most recent published work in *Red Tree Review*, *The Nation*, and *Columbia Journal*'s upcoming print issue.

Rune Graulund (BA, MA, University of Copenhagen, PhD, Goldsmiths College, University of London) is Associate Professor in American Literature and Culture at the Institute of Literature, Media and Cultural Studies, and Director of the research cluster Anthropocene Aesthetics. His research centers on American popular culture and literature (especially science fiction, clifi, gothic, and post-apocalyptic fiction) as well as political fiction and non-fiction relating to questions of migration, empire, ecology, and the Anthropocene.

Hywel M. Griffiths is a Reader in the Department of Geography and Earth Sciences at Aberystwyth University in Wales, UK, where he also studied physical geography and his PhD on the rates, processes, and controls of geomorphological processes in rivers. His main current research interests are related to memories and perceptions of extreme weather and the environment, as recorded in historical archives, including poetry, focusing mainly on Wales, but also including projects in Patagonia and Aotearoa New Zealand. He also works on science communication in collaboration with artists and creative writers.

Lesley Hawkes is an Associate Professor in the School of Communication, Faculty of Creative Industries, Education and Social Justice at Queensland University of Technology, where she is Research Training Co-ordinator for the School. Her areas of research include spatial belonging, anthropomorphism, and environmental concerns. She also has a strong research interest in the relationship between transport and imagination. She is Senior Fellow in the Higher Education Academy.

Gretchen Ernster Henderson is an intermedia writer and Senior Lecturer at The University of Texas at Austin. Her contribution seeps from her fifth book, *Life in the Tar Seeps: A Spiraling Ecology from a Dying Sea* (Trinity University Press 2023), which has been melting across publications, exhibitions, performances, and field practices. Her work has been translated across five languages. Recent publications include *Ecotone, Orion, Ploughshares*, and *LA+/Landscape Architecture Plus*. Gretchen is the 2023 Aldo & Estella Leopold Writer in Residence in New Mexico and was previously the Annie Clark Tanner Fellow in Environmental Humanities at the University of Utah. From the Sonoran Desert, in partnership with the University of Arizona Poetry Center, she is collaborating on a poetic water-harvesting project, *Dear Body of Water*.

Gabriel Kaless is a Titular Professor and Investigator at the National University of Patagonia San Juan Bosco. He teaches hydraulics in the Civil Engineering

Program, and the course 'Processes of Hydraulic Erosion' in the Specialization on Surface Water Quality. His areas of research include fluvial dynamics, morphology, sediment transportation, hydrology, and the interaction of society with hydraulic systems in the Chubut River Valley.

Alireza Korangy received his PhD from the Department of Near Eastern Languages and Civilizations, Harvard University, USA. His field of research is classical Persian and Arabic philology, with a special emphasis on poetics, rhetoric, and linguistics. He has also done extensive research on the contemporary linguistics of Iran and its corresponding folkloric traditions (Sorani Kurdish, Kurmanji Kurdish, Gilaki, Lori, Baluchi, etc.). He is Faculty at the American University of Beirut and has previously taught at the University of Virginia, the University of Colorado at Boulder, and Harvard University.

Veronika Kratz is a Postdoctoral Fellow in the Department of English at Queen's University in Kingston. She is a settler scholar researching and teaching in the environmental humanities and US literary and cultural studies. Her research focuses on narratives of environmental crisis and the cultural life of science and technology. Her work on ecology and soil conservation science in Frank Herbert's *Dune* has appeared in *ISLE*.

Elexus Lopez is a Yoeme Xicana artist. Their work focuses on intersectionality, identity, cultural awareness, and their own femme identity through different creative mediums to translate visual poetry through imagery with photography and film.Vulnerability through identity.

Tom Lynch is a Professor Emeritus in the English Department at the University of Nebraska. Lynch's primary areas of interest are ecocriticism, settler-colonialism, and place-conscious literary studies, with special attention to the American West and the Australian Outback. He has edited or authored a number of volumes, including *Xerophilia: Ecocritical Explorations in Southwestern Literature* (2008). His most recent book is *Outback and Out West: The Settler-Colonial Environmental Imaginary* (2022). For seven years, Lynch served as Editor of the journal *Western American Literature*.

Daniel Mann is a Lecturer at Queen Mary University of London and an Institute of Humanities and Social Sciences Fellow. Mann is the author of the book, *Occupying Habits* (Bloomsbury, 2022), and numerous peer-reviewed articles in journals such as *Media, Culture & Society*, *Screen*, *Media+Environment*, *Afterimage*, and *World Records*. Mann's films have been exhibited internationally at film festivals and venues such as The Berlin Film Festival, The Rotterdam Film Festival, Cinéma du Réel, and the ICA in London.

Gayatri Thanu Pillai is an Instructor in the Department of English at the National University of Singapore. Her areas of research interest are postcolonial literature and theory, South and Southeast Asian colonial literature, and ecocriticism. She has won the Maurice Baker Prize for her doctoral research on colonial South Indian literature. She is the Managing Editor ofthee *Journal of Southeast Asian Ecocriticism.*

Claudia Rosenhan works as a Lecturer at the University of Edinburgh. Her main area of interest has shifted from textual representations of race and/or gender to explore an ecocritical and posthumanist assemblage of text, environment, and translocal practices. Her most recent project was a multimodal investigation into the valley of Nethermere in D. H. Lawrence's *The White Peacock* as an ecocritical locality, focusing on energy flows. She has also written on Aldous Huxley's *Island* and E.M Forster's Cambridge Novels, with a focus on ecocritical phenomenologies.

Chitra Sankaran is an Associate Professor in the Department of English, National University of Singapore. She is the Departmental Chair of Literature. Her research interests include ecocriticism, South and Southeast Asian fiction, and feminist theory. She recently edited a special issuthe the *Journal of Ecocriticism* on 'Ecocriticism in ASEAN'. Her other publications include monographs, edited volumes, book chapters, and research articles in journals including *Material Religion*, the *Journal of Commonwealth Literature*, *Theatre Research International*, *Australian Feminist Studies*, and *Critical Asian Studies*. Her most recent monograph, Women, Subalterns, and Ecologies in South and Southeast Asian Women's Fiction, is contracted with the University of Georgia Press and will be published this year. She is the [founding] president of ASLE-ASEAN: The Association for the Study of Literature and Environment in ASEAN. She is also the Chief Ether of the *Journal of Southeast Asian Ecocriticism.*

Stephen Tooth is a Professor in the Department of Geography and Earth Sciences (DGES) at Aberystwyth University, Wales. He graduated from the University of Southampton, England, completed a PhD at the University of Wollongong, Australia, and then undertook postdoctoral work at the University of the Witwatersrand, South Africa. His main research interests focus on geography, geomorphology, and sedimentology, especially in the world's extensive drylands, which include many true desert regions. He has undertaken fieldwork in the drylands of Australia, southern Africa, Europe, South America, and the southwest USA, researching themes such as river channel-vegetation interactions, controls on gully erosion, human-environment interactions, and the use of Earth's drylands as analogues for Martian surface environments. He is also interested in environmental issues more generally, including current debates about global climate change and the Anthropocene, and in science education.

ACKNOWLEDGMENTS

The editors would first like to thank the contributors for the effort, time, and intention that they invested in this project. When we began putting the call for papers together, we weren't sure what to expect, just that we were interested in a book that would promote a positive shift by deepening knowledge about the diversity of the world's deserts and their cultures. We had begun our conversations with one another during the first winter of the pandemic, when Celina reached out to Aidan about his desert monograph which had just been published. In many ways, the timing of this collaboration coincides with a time when the world was experiencing so much collective grief and uncertainty; it seemed a hopeful thing to do to meet regularly and work on something so purpose driven. We are grateful for each other's patience, support, and commitment to doing research that is informed by the realities and vitalities of desert places.

As an early supporter of this project, we especially would like to thank Rune Graulund, whose research and encouragement has been influential to both of us. We extend our gratitude, also, to Jada Ach and Gary Reger, for their friendship and their collaborations which have brought so many together to discuss desert projects. In 2023, Aidan visited Arizona State University, where a seed grant from the Institute for Humanities Research was awarded to Lisa Han, Mako Ward, and Celina to host an interdisciplinary symposium called 'Ecologies of Justice: Wastelands, Wastewaters, and Human Disposability.' We thank Mako and Lisa for the labor and care that went into organizing and moderating this event. Celina would also like to extend her thanks to Humanities for the Environment's North American Observatory, whose sponsored panel at ASLE was invited and chaired by Joni Adamson. It was at this panel that Celina was able to present about the collection for the first time and meet contributors Veronika Kratz and Andrea Casals Hill. To Grace Harrison and Matthew Shobbrook at Routledge, we want to acknowledge that you have given a wonderful home to this project for which we will always be appreciative.

To friends, family, colleagues, and community members who have inspired and supported us in our scholarship, we extend our sincerest appreciation: Angie Bautista-Chavez, Fredrick Brown, Travis Franks, Vanessa Fonseca-Chávez, Alice Gibson, Marinela Golemi, Gia and Nosrat Heidarian, Chawa Magaña, Lydia Zulema Martinez Vega, Carlos Osuna, Jesus Sabas, and Mako Ward.

INTRODUCTION

Celina Osuna and Aidan Tynan

The premise of this book is simple: desert places and their communities are dynamic and diverse, despite dominant narratives which have homogenized 'the desert' into an abstraction. This abstraction, a nameless desert that conjures up imagery of sand dunes, sunlight, and vast horizons, has been perpetuated for centuries, crystalized by settler-colonial representations of arid environments as exotic lands where racialized "Others" or subalterns live. In the twentieth century, these aesthetics made their way into film, literature, and art that still has a profound effect on how so many understand, or misunderstand, deserts today. As a result, many deserts, as Gary Nabhan writes, are defined "by most cultures in terms of what they lack rather than by the uniqueness of what they have."[1] Perceived notions of emptiness, wasteland, and void in the popular imagination have directly affected desert places and peoples, who often suffer environmental injustices like displacement, land and water contamination, and food apartheid. Paradoxically, deserts, whether rich in precious metals or minerals, fossil fuels, or even 'open' space, have historically been and are, now more than ever, extremely valuable to governments and industries the world over. Thus, the feedback loop of representations and realities that keeps deserts in a light defined by what they and their communities do not have is in dire need of disruption. *Storied Deserts: Reimagining Global Arid Lands* contributes to the growing body of research on the world's desert places, taking an interdisciplinary, multicultural approach to the field of environmental humanities and including diverse perspectives on arid environments from around the globe. The volume discusses deserts as sites of narrative production, places where signs and imaginaries are born from the materialities of space and entanglement. In this way, we show how deserts are, despite any representations to the contrary, culturally and ecologically rich and abundant places.

DOI: 10.4324/9781003397212-1

The 'global' arid lands of our title should not be misunderstood as reinscribing the very homogenizing tendencies we are trying to counteract. This volume's international focus is intended to address one of the major challenges facing environmental humanities today, which is how to understand the relations between the local and the global. In the 1960s, the first images of Earth taken from orbit helped to inspire environmental movements by rendering up a graphic depiction of the planet as a fragile orb hanging precariously in an abyss. 'Marble Earth' tended to displace the mushroom cloud as an icon of the Western worldview.[2] But as Bruno Latour points out, global thinking risks "unifying too quickly what first needs to be composed."[3] The global is often a construct which too quickly skips over the local compositions of being placed towards a grand but illusory 'view from nowhere.' When the discipline of literary ecocriticism emerged in the early 1990s, it was moving against the grain of these global framings and their attendant universalism, preferring instead the bioregionalism of American nature writing and, subsequently, the intersectional politics enabled by convergences of the 'biocentric' and the 'sociocentric.'[4] However, the impact of the concept of the Anthropocene on both ecocriticism and the environmental humanities over the past decade has shown the insistence of the planetary view, while the relations between the Global North and South have become central to conceptions of environmental justice and the pursuit of decolonization.[5]

It is surprising how infrequently deserts have featured in these debates. Earth's drylands are local and global at once, covering some 40 percent of the planet's land surface and home to about 30 percent of the human population. But it is only very recently that scholars in environmental humanities and associated fields have begun to engage with these places in their transnational and global dimensions. In this context, desert places have generally not – with some notable exceptions – been attractive to the white Western imagination that has shaped so much of the mainstream environmental consciousness. While such imaginaries have put great emphasis on the local and the bioregional, deserts have been marginalized for being the wrong kind of place – 'badlands' and 'wastelands,' empty and dull, landscapes to be avoided or passed through as quickly as possible. Thoreau's Concord and Wordsworth's Grasmere have been seen as much more appealing prospects, more aligned with pleasant notions of Nature, as have forests and oceans.[6] When deserts have featured in the mainstream of environmental culture, they have often done so through the chauvinist and hypermasculine settler-colonial mindset of figures such as Edward Abbey, whose *Desert Solitaire* (1968) remains a revered touchstone of American nature writing despite the author's deep and unabashed racism. Even in the writings of environmentalists, 'the' desert is often presented as a homogenous and monolithic space, a pure openness or otherness into which all differences are dissolved as if in a void.

None of this is to say that our dominant cultural imaginaries are short on deserts. Indeed, the figuration of 'the' desert as a homogenous, voided other-place or 'heterotopia,' to use Foucault's influential term, has strongly informed Western settler-colonial environmental imaginaries and the kinds of violence they have wrought.[7] Certain forms of desert imagery, found particularly in American genre cinema, have become ubiquitous as a shorthand for decline, dystopia and apocalypse on the one hand and fantastical otherworldliness on the other. We need only mention here the cinematic spectacles of the *Mad Max* films and *The Road* (2009), Frank Herbert's novel *Dune* (1965) and its visually breathtaking film adaptations, or many of the films from the *Star Wars* universe. This combination of dystopian despair and sublime adventure points towards a particular aesthetics of violence and technology.

John Beck argues that the United States' desert West has been portrayed as a "site of physical and spiritual challenge, as irrecoverably hostile 'wasteland,' and as physical evidence of an achieved apocalypse" in ways that have facilitated spectacular mystifications of American military-industrial power.[8] This perfectly characterizes Christopher Nolan's Academy Award-winning film *Oppenheimer* (2023), a serious dramatization of the first nuclear weapons test in Los Alamos, New Mexico near the end of the Second World War. The film presents us with the scenic desert backdrop of classic Hollywood Westerns, with Oppenheimer himself portrayed as a brooding cowboy leading the United States to victory. The scientist-genius who holds the fate of the Earth itself in his hands is also the hero of boyhood adventure fantasies. In one of only a few times that the film mentions Indigenous people from the area, President Truman asks what should be done with the deadly, radioactive land after the testing is finished. Oppenheimer responds, in a deadpan fashion: "give it back to the Indians." The sentiment is thus suspended somewhere between a horrible joke and a serious proposal, as if the land only ever mattered as a staging ground for dreams and nightmares and now that it's become toxic and served this purpose it can be given back.

Nolan's film, though barely acknowledging it, indicates how the American military in the latter half of the twentieth century continued its war on Indigenous populations. "Nuclear colonialism began in New Mexico, where the nuclear weapons complex began," writes scholar Myrriah Gómez, who grew up in a New Mexico community directly affected by the Manhattan Project and its Los Alamos National Laboratory, describes how her "grandmother's family and other Nuevo-mexicanas/os and Tewa people were forcibly dispossessed of their ranches and sacred land on the Pajarito Plateau with inequitable or no compensation" and the subsequent contamination of land and water in the area.[9] Ian Zabarte, Principal Man of the Western Bands of the Shoshone Nation whose territory extends from Las Vegas to the Snake River in Idaho, has described his nation as 'the most bombed' on Earth due to the 900 or more nuclear weapons tests that were carried out in this desert region in the Cold War decades. "Shortly before the end of World War II," he writes, "we started to be overrun by the US military industrial complex, in ways we are only now beginning to understand."[10] Such devastation waged against the peoples and ecologies of arid lands can be investigated through the influence of an environmental aesthetics in which deserts are systematically devalued via an imperialist spatiality.

Indigenous writers such as Ofelia Zepeda (Tohono O'odham) and Leslie Marmon Silko (Laguna Pueblo, Mexican and Anglo-American heritage) in North America and Alexis Wright (Waanyi) in Australia, to name just a few, have done much to reclaim the significance of desert places, celebrating cosmologies and constructing imaginaries in which a plenitude of cultural traditions and personal histories displace and resist the typical Western stereotypes of howling void and unpeopled wilderness. One of the major goals of this volume is to interrogate, critique and ultimately counter dangerous misrepresentations of desert places, exposing them to be the political constructs that they are. The failure of scientists to come up with a single formal definition capable of encompassing the multifariousness of desert places is notable in this regard.[11] This apparently singular category of landscape can contain multitudes in ways which dominant environmental imaginaries often fail to recognize.

We also wish to dispel a sense that deserts can be reduced simplistically to 'natural' or inert materialities, or to mysterious spaces of geographical alterity. Deserts may harbor mysteries, of course, but we should resist tendencies to exoticize them, as the West's orientalizing imagination has so often done. Such a shift involves recognizing that desert places are populated through and through, abundant in lived realities, ecological entanglements, cultures, histories, and futures. Reducing this rich diversity to a bare natural or inert material fact named 'the desert' is committing a deep injustice. To this end, recent developments in ecological theorizing in the humanities and social sciences suggest not only a dissolution of nature/culture binaries but also a tendency to recognize matter as agential and vibrant.[12] For work in ecocriticism specifically, this has led to conceptions of matter as 'storied,' imbued with meaning and agency. As Iovino and Oppermann write in their introduction to the edited volume *Material Ecocriticism*, "matter and meaning constitute the fabric of our storied world" in ways that render dualisms of nature and culture, nonhuman and human, obsolete.[13] Our volume takes inspiration from these ideas. Though the notion of 'storied deserts' stems from Osuna's *Desert Distortion*, a monograph on desert abundance in the so-called Desert Southwest, as this project took shape, the notion developed to become a generative way of engaging with arid environments around the world – their communities, histories, politics, art, literature, and futures. Thus this book aims to attest to the truly diverse nature cultures of global arid lands; it is not bound by region, timeline, or discipline and instead seeks to reflect a heterogeneity of experiences, with each author contributing their understanding of deserts as storied, agential, complex places.

While the neglect of deserts has been a notable feature of environmental humanities scholarship, there are key existing works upon which this project builds. We would like to acknowledge prior and emerging work in desert humanities. Monographs that refute ingrained biases about what ecologically significant places are supposed to look like include Scott Slovic's *Getting Over the Color Green* (2001), Catrin Gersdorf's *The Poetics and Politics of the Desert* (2008), and Tom Lynch's *Xerophilia* (2008). These books demonstrate the importance of shifting dominant, and often harmful, attitudes about desert places. Lynch makes the crucial point that when it comes to writing on deserts, a vocabulary of lack and deficiency has become ingrained in the English language, even in the case of desert-loving writers such as Mary Austin.[14] A xerophilic writing – that is, a writing adapted to the arid or the dry, the xēros – would provide us with an entirely different bioregional imagination than the one founded on the shores of Walden Pond. More recently, monographs such as Tynan's *The Desert in Modern Literature and Philosophy: Wasteland Aesthetics* (2020) and Jada Ach's *Sand, Water, Salt: Managing the Elements in Literature of the American West, 1880–1925* (2021) have been published, the former interested in desert imaginaries and the latter investigating how these imaginaries do or don't translate to land management. Osuna's *Desert Distortion* and Gary Reger's *Wild, Weird, West: Essays on Arid America* are due out soon, and both texts take on the complexities of desert histories and narratives with special attention to the long twentieth century.

It is clear to us, as scholars who have been keenly interested in deserts and producing work to investigate their aesthetics and representations in literature, art, philosophy, and film, that a focus on deserts across disciplines is taking shape. In the four-year process of putting this volume together, there have been at least as many edited collections published – each of which is interested in doing away with the notion of the desert as void in order to instead understand deserts and their communities as diverse and important. Both *Reading Aridity in Western American Literature* (2020) and *The Invention of the American Desert* (2021) express a desire for a deeper engagement with desert places in the American West and beyond from interdisciplinary perspectives. *Deserts Are Not Empty* (2022) and *Desertscapes in the Global South and Beyond* (2023) are further crucial contributions to the field and, like *Storied Deserts*, provide more planetary approaches informed by local engagements that decenter Western perspectives, what Samia Henni calls "the same old colonial platitudes" about equating deserts with emptiness, in ways that are directly political.[15] Interestingly, both *Deserts Are Not Empty* and *The Invention of the American Desert* illuminate architectural interests in deserts in the twenty-first century, clearly a key issue for a world in which existing urban infrastructures are proving inadequate in the face of climate crisis. Henni explains, "architecture is not only what is designed and built but also what is indebted, destroyed, dismantled, contaminated, displaced, buried, unearthed, and wasted."[16] It is, of course, the populations of the planet's arid regions, mostly living in the Global South, who are most exposed to the effects of increasing temperatures and a collapsing climate system. Reimagining deserts is thus a crucial task for surviving the Anthropocene and envisioning more just futures.

At the same time, we should not ignore how the global character of climate crisis tends to reinscribe homogenizing images of deserts as growing wastelands. As Veronika Kratz's chapter in this volume explores, what has been called, misleadingly, 'desertification' conjures up a vision of encroaching sands swallowing cities or whole civilizations in ways that obscure the real agricultural and economic complexities while transferring blame for ecological devastation from the most powerful to the least. As Diana K. Davis points out in her work on the arid lands of North Africa, the idea and term 'desertification' arose out of nineteenth-century French colonial discourse and was used to blame nomadic Indigenous peoples for soil erosion.[17] Hannah Hollemann offers 'dustbowlification' as a more accurate term in this respect, linking contemporary soil erosion directly to the forms of settler-colonial agriculture that were a causal factor in the US Dust Bowl disaster of the 1930s.[18] When a writer and activist as astute as Mike Davis[19] describes the Anthropocene in terms of "the coming desert," it shows the tenacity of what Tynan has called "wasteland aesthetics," an environmental imaginary of destitution and devastation that has come to characterize Western ideas of spatiality since at least the end of the eighteenth century.[20] Conceptions of the desert as a wasteland emerged with particular force around this time partly as a result of a new aesthetic awareness of landscape, influenced by Romanticism, but partly because of the encounter with spatial alterity produced by colonial expansion. The desert as wasteland thus took on a profound meaning for a European subjectivity contending with its own reality as an actor of planetary extent.

Desert humanities scholarship thus needs to counter conflations of deserts and wastelands while also understanding the persistently oppressive power of such conflation. Elizabeth Povinelli, an anthropologist who has worked extensively with the Indigenous people of Australia's Northern Territories, writes that "the figure of the Desert" is a key part of contemporary global power structures whose impact is particularly felt in place-based struggles over land and resources.[21] As Traci Brynne Voyles writes in a history of uranium mining on Diné (Navajo) land,

> Notions of Navajo country as 'uninhabited' wasteland create a representational criterion by which ideas about the land have been formed. … The land, occupied and claimed by tribes, with its own unique sets of ecological conditions and realities, ceased to be an empirical object – the material conditions of Narbona Pass, with its shimmering greens and crisp air, is forgotten in favor of an interpellation of Navajo country writ large as wasteland.[22]

Western imaginaries posit the desert as a figure of non-life, a place where life has been or may be again, an inert substance subject to extractive violence, or a site of waste and/or disposal. Furthermore, these Western notions of deserts as void and non-living have made their impacts known across the Global South. In an example akin to the one above, claims that deserts are devoid of life are what the French army used "to defend its choice to use the Algerian Sahara – then a French colonized territory – as the firing field for France's first atomic bombs between 1960 and 1966."[23] At the same time, the desert has been imagined in various different traditions as a site of escape, transgression, alterity, healing, and becoming by writers, filmmakers, and environmentalists whose ideas are sometimes, but by no means always, in dialogue with Indigenous peoples for whom desert places are home. How do we traverse these different geographical and cultural spaces? What is at stake in both conceptual and political contestations of desert lands? What, ultimately, does this tell us about relations between belonging, territory, its loss, or return? The borders between the aesthetic and the political here are porous and shifting.

The world's arid lands are generally understood in terms of their scarcities or limiting factors, but this book's aim is to reimagine them according to the multifarious modes of human and nonhuman life they enact and enable. *Storied Deserts* (re)establishes these places as abundant and diverse, even as some chapters explore the historical, cultural, and sociopolitical negation of these things. The idea of storied deserts is one that works not only to overturn long-established figures of the desert as empty wasteland, it is also meant to move traditions, histories, and cultures that have been systematically subjugated or erased to the forefront of our environmental consciousness. This task is seemingly simple in concept, but endlessly complex in practice. It needs us to attend to the immense power of environmental imaginaries – their capacity to value or devalue certain places, along with their inhabitants, through symbols, narratives, images, and affects – while articulating a generative vision of deserts as productive of stories capable of engaging the past, addressing present injustices and cultural community values, and articulating and then working towards desirable, robust futures. These are the main goals orienting the interdisciplinary, transhistorical, and geographically wide-ranging work collected in this volume, the work of storied deserts.

Book Structure

The book is divided into three parts, beginning with innovative interpretations of familiar portrayals of desert places, moving through engagements with traditions of poetry from different cultures, and eventually grappling with geopolitical and cultural complexities of belonging to and even being displaced from arid lands. Part I: Elemental Narratives provides critical analyses of how deserts have been represented in the Western cultural imagination, ranging from popular cinema to literary fiction and philosophy, as sites of elemental intensity defined by forces such as heat, light, and radioactivity. Part II: Poetics of Aridity gives prominence to desert poetry from and about the Global South, presenting arid lands as generative of semiotic abundance and featuring some creative works from a practicing artist. And lastly, Part III: Land: Politics of Belonging challenges long-established narratives of desert as void or damaged land to assert place-based narratives of identity and justice and includes the volume's second artistic contribution.

In his chapter, scholar and director Daniel Mann offers a critique of the desert as a favored location for Hollywood narrative. Mann opens with a discussion of an image of iconic actor John Wayne on the set of the 1955 film *The Conquerer*, shot on irradiated land near one of the most significant of America's nuclear weapons test sites. He argues that American cinema's use of deserts around the world, like in Jordan and Morocco, expresses a colonial gaze. Rune Graulund keeps with the theme of popular genre narratives, specifically science fiction, but shifts focus to the Antarctic – the world's largest desert – offering an intellectual history of the driest, coldest continent. While the Antarctic landscape is rarely addressed in terms of desert imaginaries, its elemental extremes have played, as Graulund shows, a key role in Western figurations of void and otherworldly space familiar in films like *The Thing* (1982) and *Godzilla vs. Kong* (2021). Delving further into Western figurations of desert as landscapes of fear, Lesley Hawkes discusses Australia's complex relationships with its arid lands. To do so, Hawkes draws on a number of literary and cinematic works to explore contrasting visions of doom and potential, moving from colonial narratives and the dystopias of *Mad Max* to Indigenous histories and more hopeful, reparative accounts. Claudia Rosenhan's chapter rounds out the first part with a piece that emphasizes light as key to desert elementality in canonical desert literature like Don DeLillo's *Point Omega* (2010) and Michael Ondaatje's *The English Patient* (1992). Building on the philosophical insights of Deleuze and Guattari and deploying Karen Barad's conceptions of 'spacetimemattering,' Rosenhan reads a series of novelistic portrayals of luminosity, diffraction, and radiant energies.

Moving towards perspectives and poetic texts from arid lands in the Global South, the volume's second part opens with Chitra Sankaran and Gayatri Thanu Pillai's ecofeminist readings of contemporary Tamil poetry, fiction, and film from India and Sri Lanka. Sankaran and Pillai draw on Tamil poetic traditions to challenge, but also transform, wasteland imagery and to explore entanglements of land, agency, and the politics of sex and gender. An exploration of Alireza Korangy's chapter provides readings of a number of poets from pre-Islamic, or Jāhiliya, Arabia, emphasizing human-animal relations, romance, and forms of verse such as the *ghazal*. Korangy shows how nomadic movements – the rhythms of a camel's feet, for example – can come to inform poetic creativity. Pursuing the relationships between poetry and landscape, Hywel M. Griffiths, Stephen Tooth, and Gabriel Kaless combine perspectives from physical geography and hydraulic engineering with literary history and autoethnographic reflection. This chapter provides personal accounts from field research trips to the semi-arid lands of Argentine Patagonia. It also documents the experiences of nineteenth-century Welsh colonists in the region to highlight how '*y Paith*' – the desert – became a focus of Welsh poetry and cultural memory. And finally, this section of the volume ends with poetry, photographs, and personal reflections from Gretchen E. Henderson, whose experimental work takes inspiration from Robert Smithson's famous piece of land art, assembled on a shore of Utah's Great Salt Lake, *Spiral Jetty* (1970). Henderson concentrates on the region's tar seeps to meditate on the agencies and dynamisms of this so-called 'dead' environment.

The final part of the book opens with Veronika Kratz's critical interrogation of narratives which perpetuate even as they misunderstand desertification, which the author clarifies is a term for the erosion or degradation of soil. Challenging established conflations of desert with damaged land, Kratz shows how discourses of desertification emerged in the 1970s from colonial governments in Africa and have been subsequently deployed in neocolonial manifestations of Western power. Shifting focus to the Atacama Desert – the world's driest hot desert – in northern Chile, authors Andrea Casals Hill and Pablo Chiuminatto cover a wide range of literary authors, and center women's voices, to show how the Atacama is rich in signs and stories, some of which pertain to the bloody and traumatic events from the nation's past. Ecocritic Tom Lynch, whose work on deserts is foundational to a growing field of desert humanities, returns us to Australia with a chapter focused on the management of Uluṟu-Kata Tjuṯa National Park (formerly Ayers Rock), perhaps that nation's most iconic geographical site. Lynch mixes personal reflection and political commentary on the Uluṟu handback, where the land has been returned to the Aṉangu people to jointly manage, and connects this to Indigenous Land Back movements in the United States. Celina Osuna's chapter analyzes contemporary works of fiction and poetry from the US-Mexico borderlands, deploying her storied desert approach as a means of encompassing environmental elementality and the politics of migration. Osuna emphasizes abundance as a means for rethinking life in desert places, foregrounding the roles of language and community. This final part of the volume concludes with a poem by Anna Flores and a series of black and white photographs from Elexus Lopez. Taken together, these pieces comprise a collaborative vision highlighting the shared experiences of erosion and endurance and traversing various kinds of borders to connect flesh and rock, the living and the dead.

Throughout this collection, there is a balance of contributions that are interested in the influences of desert aesthetics and those that reveal lived realities, whether historical or contemporary, of actual desert places. Some chapters strike this balance within themselves. Brought together here, it is our hope that they illuminate the generative approach of storied deserts – one that requires us to shift perceptions away from desert places as barren and towards a deep engagement with their abundance. A reimagining of the world's arid lands, which have most commonly been portrayed through settler-colonial narratives interested in extractive economies and exoticism, is overdue, especially as understanding the biological and cultural diversity of deserts is key to sustainable human futures and planetary health. It is in this light that we offer *Storied Deserts: Reimagining Global Arid Lands* as a project that takes seriously the importance of stories and storied matter alike, as deserts have been too often misrepresented as devoid of both. Ultimately, we hope that readers will enjoy the array of perspectives and possibilities presented here – scholarship that diffracts 'the desert' into a vast range of storied deserts.

Notes

1 Nabhan, *The Nature of Desert Nature*, 14.
2 Diederichsen and Franke, *The Whole Earth.*
3 Latour, *Facing Gaia*, 138.
4 Garrard, "Ecocriticism," 62.
5 Adamson, "Environmental Justice, Cosmopolitics, and Climate Change."

6 See Harrison, *Forests*; Kohn, *How Forests Think*; Alaimo, "Introduction"; Mentz, *An Introduction to the Blue Humanities*.
7 Foucault, "Of Other Spaces."
8 Beck, *Dirty Wars*, 22.
9 Gómez, *Nuclear Nuevo Mexico*, 5.
10 Zabarte, "A Message from the Most Bombed Nation on Earth."
11 Welland, *The Desert*, 17.
12 Bennett, *Vibrant Matter: A Political Ecology of Things*.
13 Iovino and Oppermann, *Material Ecocriticism*.
14 Lynch, *Xerophilia*, 34.
15 Henni, *Deserts Are Not Empty*, 12.
16 Henni, 20.
17 Davis, "Desert 'Wastes' of the Maghreb."
18 Holleman, *Dust Bowls of Empire*.
19 Davis, "The Coming Desert: Kropotkin, Mars and the Pulse of Asia."
20 Tynan, *The Desert in Modern Literature and Philosophy*.
21 Povinelli, *Geontologies*.
22 Voyles, *Wastelanding*, 8.
23 Henni, *Deserts Are Not Empty*, 11.

Bibliography

Ach, Jada. *Sand, Water, Salt: Managing the Elements in Literature of the American West, 1880–1925*. Texas Tech University Press, 2021.

Ach, Jada, and Gary Reger, eds. *Reading Aridity in Western American Literature*. Ecocritical Theory and Practice. Lexington Books, an imprint of The Rowman & Littlefield Publishing Group, Inc., 2020.

Adamson, Joni. "Environmental Justice, Cosmopolitics, and Climate Change." In *The Cambridge Companion to Literature and the Environment*, edited by Louise Westling, 169–183. Cambridge Companions to Literature. Cambridge University Press, 2013. https://doi.org/10.1017/CCO9781139342728.016.

Alaimo, Stacy. "Introduction: Science Studies and the Blue Humanities." *Configurations* 27, no. 4 (2019): 429–432. https://doi.org/10.1353/con.2019.0028.

Beck, John. *Dirty Wars: Landscape, Power, and Waste in Western American Literature*. University of Nebraska Press, 2009.

Bennett, Jane. *Vibrant Matter: A Political Ecology of Things*. A John Hope Franklin Center Book. Duke University Press, 2010. https://doi.org/10.1515/9780822391623.

Davis, Diana K. "Desert 'Wastes' of the Maghreb: Desertification Narratives in French Colonial Environmental History of North Africa." *Cultural Geographies* 11, no. 4 (October 2004): 359–387. https://doi.org/10.1191/1474474004eu313oa.

Davis, Mike. "The Coming Desert: Kropotkin, Mars and the Pulse of Asia." *New Left Review* 97 (February 2016): 23–43.

Diederichsen, Diedrich, and Anselm Franke, eds. *The Whole Earth: California and the Disappearance of the Outside*. Translated by James Gussen. 2. ed. Sternberg Press, 2013.

Foucault, Michel. "Of Other Spaces." Translated by Jay Miskowiec. *Diacritics* 16, no. 1 (1986): 22–27. https://doi.org/10.2307/464648.

Garrard, Greg. "Ecocriticism." In *Keywords for Environmental Studies*, edited by Joni Adamson, William A. Gleason, and David N. Pellow, 61–63. Keywords 3. New York University Press, 2016. https://doi.org/10.18574/9780814760741.

Gersdorf, Catrin. *The Poetics and Politics of the Desert: Landscape and the Construction of America*. Brill Academic Pub, 2009.
Gómez, Myrriah. *Nuclear Nuevo México: Colonialism and the Effects of the Nuclear Industrial Complex on Nuevomexicanos*. University of Arizona Press, 2022.
Harrison, Robert Pogue. *Forests: The Shadow of Civilization*. Paperback ed., [Nachdr.]. University of Chicago Press, 2009.
Henni, Samia, ed. *Deserts Are Not Empty*. Columbia Books on Architecture and the City. Graduate School of Architecture, 2022.
Holleman, Hannah. *Dust Bowls of Empire: Imperialism, Environmental Politics, and the Injustice of "Green" Capitalism*. Yale University Press, 2018.
Iovino, Serenella, and Serpil Oppermann. *Material Ecocriticism*. Indiana University Press, 2014.
Kohn, Eduardo. *How Forests Think: Toward an Anthropology Beyond the Human*. University of California Press, 2013.
Latour, Bruno. *Facing Gaia: Eight Lectures on the New Climatic Regime*. Translated by Catherine Porter. First published. Polity, 2017.
Lynch, Tom. *Xerophilia: Ecocritical Explorations in Southwestern Literature*. Texas Tech University Press, 2008.
Massey, Lyle, and James Nisbet, eds. *The Invention of the American Desert: Art, Land, and the Politics of Environment*. University of California Press, 2021.
Mentz, Steve. *An Introduction to the Blue Humanities*. Routledge, Taylor & Francis Group, 2024.
Nabhan, Gary Paul, ed. *The Nature of Desert Nature*. Southwest Center Series. The University of Arizona Press, 2020.
Osuna, Celina. *Desert Distortion: Revealing Desert Abundance*. Texas Tech University Press, 2025.
Povinelli, Elizabeth A. *Geontologies: A Requiem to Late Liberalism*. Illustrated edition. Duke University Press Books, 2016.
Reger, Gary. *Wild, Weird, West: Essays on Arid America*. Texas Tech University Press, 2024.
Shekhawat, Sushila, Rayson K. Alex, Swarnalatha Rangarajan, and Scott Slovic, eds. *Desertscapes in the Global South and Beyond: Anthropocene Naturecultures*. Routledge Studies in World Literatures and the Environment. Routledge, 2024.
Slovic, Scott. *Getting over the Color Green: Contemporary Environmental Literature of the Southwest*. University of Arizona Press, 2001.
Tynan, Aidan. *The Desert in Modern Literature and Philosophy: Wasteland Aesthetics*. Crosscurrents. University Press, 2020. https://doi.org/10.1515/9781474443371.
Voyles, Traci Brynne. *Wastelanding: Legacies of Uranium Mining in Navajo Country*. 1st edition. University of Minnesota Press, 2015.
Welland, Michael, ed. *The Desert: Lands of Lost Borders*. Reaktion Books, 2015.
Zabarte, Ian. "A Message from the Most Bombed Nation on Earth." *Al Jazeera*, August 29, 2020. https://www.aljazeera.com/opinions/2020/8/29/a-message-from-the-most-bombed-nation-on-earth.

PART 1

Elemental Narratives

1

A DUEL IN THE SAND

On Cinematic Spectacle in Arid Lands

Daniel Mann

FIGURE 1.1 John Wayne on the set of The Conqueror with a Geiger counter. The Geiger counter clicked, indicating the radioactivity in the ground below. Photograph: Supplied by *The Guardian*, https://www.theguardian.com/film/2015/jun/06/downwinders-nuclear-fallout-hollywood-john-wayne 'Hollywood and the downwinders still grapple with nuclear fallout', 6 June 2015.

DOI: 10.4324/9781003397212-3

Looking at the above photograph, one will likely recognise the familiar face of John Wayne. Less familiar are Wayne's two sons, Michael and Patrick, who both played uncredited minor roles in the blockbuster film, *The Conqueror* (Dick Powell, 1955), filmed in Utah in 1954. The picture was taken on set during a break in the filming. Wayne is shown here grasping an otherwise non-descript black box as his sons look on intently, and a fourth man appears to be adjusting something on the box itself. What the photograph does not capture is the sound emanating from the box—a Geiger counter, as it happens—nor the fact that it was crackling so loudly that Wayne thought it must be defective. It wasn't. No matter where it was directed, across different clumps of rock and sandy soil in the environs near the set, the result was the same. More to the point, the Geiger counter went off despite the fact that the desert film location was more than one hundred miles from the Nevada Test Site established in 1951 by the US government to conduct tests on the country's rapidly expanding nuclear stockpile. As would later emerge when nearly half the crew, including the director, were diagnosed with various cancers, mostly well below average age, those on the set of The Conqueror were possible unwitting 'downwinders', exposed to radioactivity from the above-ground testing of nuclear weapons through the early 1950s (Figure 1.1).

Today scientists understand the extent of the scope of the threats posed by the above-ground testing of nuclear weapons in the region. At the time, the Nevada Test Site (later renamed Nevada National Security Site in 2010) was chosen as a 'safe' site to conduct this kind of testing since it was assumed that winds would blow 'radiological hazards' away from the population centres of Las Vegas and Los Angeles towards 'virtually uninhabitable' land downwind to the west.[1] It is known that more than 100 nuclear detonations occurred between 1951 and 1962, sending huge toxic plumes of radioactive dust across the stony valleys and canyons of southern Utah and northern Arizona, some of Hollywood's most popular film locations. The official who recommended the site for the NTS was Navy Captain Howard B. Hutchinson, a 'highly qualified meteorologist'. Following the end of the Second World War, Hutchinson was put in charge of evaluating the 'physical feasibility' of conducting nuclear weapons tests within the continental United States.[2] Hutchinson's brief was to determine 'how, when, and where' tests could be conducted without radioactive fallout causing 'physical or economic detriment to the population'. He narrowed his analysis down to the arid southwest and the humid southeast. The southwest, he decided, was 'more favourable' for 'purposes of planning and logistics'.[3] The desert, he concluded, was remote, uninhabited and not fit for any productive purpose. 'Few areas in the continent are more ruggedly severe and inhospitable for humans. Only once prior to 1950, and then very briefly, did more than a hundred people call the place home', noted a report issued by the United States Department of Energy.

The photograph from the set of *The Conqueror* only hints at the radioactivity absorbed into the rocky terrain of the Utah desert, chosen, of course, for its suitability for epic depiction of Westerns and desert warfare. Here, the appearance of the desert assumes a dual character. In the first place, we observe a barren, virginal, unspoiled desert backdrop, which repeatedly features as a two-dimensional image behind the action in hundreds of big-budget Hollywood blockbusters. But on the other side, we observe the topography of the desert, silently absorbing the blows of humanity's insatiable drive to technological dominance, at once an integral part of the continental United States and at the same time alienated from it, a timeless landscape observing the long sweep of human history. The former appears as a horizontal surface image while the latter is localised, historical and vertical.

The photograph serves as a reminder of the enduring relationship between high-budget Hollywood cinema and desert environments. This relationship is at once spectacularly visible through the thousands of films shot in the desert and at the same time hidden from view, deep below the surface. Desert topography, as the photograph suggests, is not only 'empty frontier space' to be exploited to conceal or indeed showcase the fruits of humanity's most deadly technology. Open, arid, sunlit, vast and so often perceived as vacant, unusable, remote and easily controlled by armies and governments, deserts have long served in Hollywood film productions as backdrops for narratives about life on national and planetary frontiers, where resource scarcity spirals into violent wars.

To counter this notion of wasteland, where cinematic fiction and spectacle seem to be conveniently produced, we must today undo the Euro-American desert imaginaries to reenvisage their position in our collective perception of Earth. Against the tendency of big-budget cinema to use the desert as a mere backdrop for anthropogenic narratives, we need visualisations of deserts that can resist the violent abstraction sustained through colonial and neo-colonial lenses. More than any other media, Hollywood cinema disseminates and shapes a collective perception of a fictional, abstracted and extra-terrestrial desert.

From 1910 onwards, the rapidly growing Hollywood film industry marked out deserts as a particularly suitable landscape for filmmaking. Hollywood's attention reinforced the national imaginary of the western United States as an untamed and unspoiled wilderness. At the same time, the desert environment offered Hollywood material properties uniquely suited to the needs of the industry, including long hours of daylight, accessibility, and, most importantly, barren land that could readily serve as a backdrop for the widest range of narratives. Cinema has been inseparable from the process of making the desert of the southwest familiar to American and later international audiences while at the same time perpetuating the systematic blurring of desert places and the human and nonhuman life within. A unique and paradoxical optical logic thus links spectacular cinema and deserts, through which the cameras of big-budget films capture arid lands as representations while simultaneously partaking in a much larger process of abstraction that extends the colonial perception of terra nullius.

But in the last few decades, Hollywood cinema has moved on from the American southwest. For the locations that once served Hollywood well in films depicting frontier life or imagined worlds were becoming ever more familiar to the American filmgoer. With its iconic topographical formations, the unmistakable landscapes of the American Southwest became somehow prosaic, fuelling a search for new landscapes less familiar to the Western eye. The familiarity of locations in California, New Mexico, Arizona, Nevada and Utah, together with the rising costs of insurance and the increasingly stringent requirements to obtain permission to film in the national parks and federal reserves in which most deserts are contained, encouraged Hollywood producers to look elsewhere. To that end, Hollywood has ventured out to areas well beyond the United States in search of new terrain for use as a backdrop in science-fiction and war films.

From the 1970s onwards, governments in the Middle East and North Africa began to offer tantalising solutions to many of the challenges Hollywood was trying to resolve, each with its own set of incentives configured in a distinctive neo-colonial package. Jordan, Morocco, the United Arab Emirates (UAE) and Israel have all offered various kinds of regulatory gives at different times, including suspension of land use regulations, tax rebates and other financial incentives, and militarised spaces to smooth the path for Hollywood producers seeking desert landscapes to depict imagined pro-filmic worlds. *Dune: Part One* (2021, directed by Denis Villeneuve), *Mission to Mars* (2000, Brian De Palma), *The Martian* (2015, Ridley Scott), *Last Days on Mars* (2013, Ruairi Robinson), *The Red Planet* (2010, Antony Hoffman), *Transformers* (2007, Michael Bay), Rambo (1987, Peter MacDonald), *Game of Thrones* (2011–2019, David Benioff and D.B Weiss), *Star Wars* (1977, George Lucas), *Red Planet* (Antony Hofmann, 2000), *Star Trek Beyond* (2016, Justin Lin), and *Fast & Furious* (2015, James Won) are only a few of the many multi-million-dollar runaway productions shot in the Sahara, Arabian, and Naqab deserts. The line that runs through this variegated list of productions, and dozens of others, is the use of the environment as infrastructure in high-budget visual entertainment, where the imagination of planetary frontiers is captured on film.

Instead of focusing on representations in films and perpetuating the visualisation of the desert as an allegorical space, my intention in this chapter is to look more closely at desert backdrops and probe their materiality for alternative stories. In so doing, I wish to touch on one of the main interrogations of this volume through questioning how matter becomes, in Serenella Iovino and Serpil Oppermann words: ‘a site of narrativity, a storied matter, a corporeal palimpsest in which stories are inscribed’.[4] The materiality of the desert poses a counter-narrative to the level of the representation. It demands a different ethics, one that invites entanglement with other human subjects and non-human objects, generating strange stories that complicate what is seen in the frame. Allowing such ‘storied matter’ to intervene with the cinematic spectacle and depictions of otherworldly space means pulling them ‘down to earth’ to uncover the material relations (i.e., the local geopolitics of land dispossession, free economic zones and geological specificity) that hold together the production of science-fiction television and film. While much has been written on the representation of the desert in film and its appearance as a backdrop for narratives about the ‘frontier’, cinema’s reliance on hyper-arid lands as material infrastructure for the production of films is underexamined. By displacing attention from the plots of films to the ground that supports them, this chapter aims to address the violence intrinsic to the abstraction of the Earth. Furthermore, stripping out the colonial imaginary of the desert as a denuded and lifeless land reveals a neo-colonial economy that makes use of unique regulatory and economic benefits set by local governments to lure the rich screen-based industry into their territory, where film producers use the desert as a playground for the production of fiction.

Duel in the Quicksand

The French philosopher Michel Serres opens his book *The Natural Contract* (1990) with a description of the barren landscape in which the two men skirmishing in Francesco Goya's 1820 painting titled 'Fight with Cudgels' are surrounded (Figure 1.2).[5] The duelling figures , one whose head appears to be bleeding, are captured knee-deep in the quicksand, cudgels raised, ready to strike the mortal blow. The drama of the moment is reflected in the bleeding face, weapons poised mid-swing and the speculation as to which of the men will prevail. But Serres enjoins his readers to resist the temptation to venture on the outcome of the duel and instead to direct their gaze at the environment in which the fighters are ensconced. There, notes Serres, in the dead boggy marsh surrounding the men, another story is unfolding:

> The painter, Goya, has plunged the duellists knee-deep in the mud. With every move they make, a slimy hole swallows them up, so that they are gradually burying themselves together.[6]

As we concentrate our attention on the deadly intent of these two combatants, each attempting to strike the mortal blow against the other, we scarcely notice that the very ground of their struggle threatens to swallow them both as they steadily sink

FIGURE 1.2 'Fight with Cudgels', Francisco Goya, 1820–1823. Alamy Stock Photo.

into the quicksand. Through this painting, Serres wishes to challenge the contours of the rehearsed irreconcilable split between the human action that catches the eye of the observer and the material environment that ensconces that action, a division which recalls the foreground–background divide in the world of theatre, where the background setting merely holds together the human actors in the foreground of a scene. It is this divide that Serres wishes to contest by redirecting our attention to the ground into which the duellists—and also we, the voyeurs—are sinking. He asks: 'in what quicksand are we, active adversaries and sick voyeurs, floundering side by side?'[7]

There is a faintly cinematic resonance in Serres' intervention. On the one hand, cinema must position its characters in a setting from which they can act and live in the world. However, unlike a painting, no film can ever be fully eviscerated from the material background it uses as a backdrop, even if the film's location is carefully obscured. Something connects the constructed world to the native geography, an affinity that exceeds the relation of representation. This so-called affinity, for German theorist and publicist Siegfried Kracauer, is not film's association with any naive authenticity or verisimilitude but with the ability to rediscover and articulate materiality through and with film, to enact 'the process of materialization'.[8] Kracauer believed that film and its connection to physical reality introduced 'a historical turn to materiality' because it confronts facticity and contingency through the imprint of the world on the celluloid. Film sees too much, or in his words, 'film looks under the table', deflating myths and conventions dislocated from their material basis and revealing hidden layers.[9] Fabricated as some pro-filmic worlds might be, film cannot fully divest itself from the place it represents, from the location in which it is produced. Kracauer, not unlike Serres, recognised the world of things as teeming with life and the materiality of film as intimately related to the topography of the represented land.

It is in the desert that cinema continues its attempt at ridding itself of this inextricable link to placeness and topography. Vast dunes, canyons and salt plains are seemingly unrecognisable to the viewers' eyes and therefore assist the creation of worlds that are divorced from their native geography. Or so Hollywood hoped. But the photographed land keeps betraying its real locality. Elena Gorfinkel and John David Rhodes stress that the specificity of any place will be both captured and alluded to by moving image media: 'Set up a movie camera in front of a place … and it will be both exactly what it records and exactly what it cannot bear witness to'.[10]

A cinematic frame carries a fragment from a particular place, but it cannot impart to us the whole place. This means, paradoxically, that while cinema represents places, it simultaneously hides them, converting them into backdrops for anthropogenic action. The very metaphor of the 'backdrop', Gorfinkel and Rhodes write, resonates with the typical figure–ground relations that obtain in the construction and consumption of moving images: 'very often, the place of the image is registered as mere "backdrop"—the necessary material support for the human action that will transpire before it'.[11] But, just like in Goya's painting, sometimes that material support gives way and begins to pull the characters down to earth. Isn't that what Serres meant in his injunction that we pay attention to the ground that is steadily consuming Goya's duellists? 'Take away the world around the battles', he writes, 'keep only conflicts or debates, thick with humanity and purified of things, and you obtain stage theatre, most of our narratives and philosophies, history, and all of social science: the interesting spectacle they call cultural'.[12] Desert locations have often been adopted as spaces stripped of place, as stage theatres for narratives of colonial expansion.[13]

Duel in Death Valley

One hundred years after Goya painted his duel, German émigré Erich von Stroheim directed the silent epic *Greed* (1924), which, in its way, recalls the painter's vision of man set against nature. Moreover, shot entirely on location in various parts of California, von Stroheim's film marked Hollywood's 'arrival' in the Golden State, especially in its use of local environments, particularly California's vast, dry desert spaces. Based on Frank Norris' novel *McTeague* (1899), Von Stroheim's film tells the cautionary tale of two estranged friends, McTeague, whom everyone calls 'Mac' (Gibson Gowland) and Marcus (Jean Hersholt), undone by a fateful lottery win that triggers a chain of events driven by greed and jealousy.[14] The plot revolves around a $5,000 lottery prize won by Mac's wife, Trina (who is Marcus's cousin), which results in a rift between the two men and culminates in Mac murdering Trina and pocketing the loot before escaping into the desert on horseback. Feeling cheated, Marcus pursues him in the hope of seizing the money for himself. The film's climatic sequence depicts Marcus' pursuit of Mac through the flatlands of Death Valley in the blistering mid-summer heat.

We watch a fatigued and badly sunburnt McTeague as he leads his horse over the cracked earth, the blazing sun overhead. By the time Marcus, himself beat and battered by the scorching heat, finally catches up with Mac, one of the horses has already succumbed to the desert, and water is running low. Surrounded by the unforgiving desert terrain, the two face each other in a duel to the death (Figure 1.3). Finally, after a protracted battle in the salt flatlands of the Death Valley, McTeague bludgeons his old friend-turned-foe to death —only to realise that Marcus has handcuffed them together during the struggle. The film ends with McTeague, desperate and worn out, chained to a corpse, with no water, left to die alone under a scorching sun.

While the film's plot narrates a cautionary tale about the corrupting effects of avarice and envy, *Greed*'s production tells a different story, about the perils of shooting a motion picture in the harsh terrain of California's Death Valley. A pioneer in cinema verité, Greed was the first feature film ever shot entirely on location, without any studio set involved. Von Stroheim also leveraged the environment to draw bracingly real emotion from his actors, leading his crew deep into the heart of Death Valley in mid-August to film the closing scenes. Goldwyn studio executives

FIGURE 1.3 Still from *Greed* (Erich von Stroheim, 1914). Alamy Stock Photo.

sought to dissuade him, and their insurance company warned of arsenic-poisoned water holes, treacherous quicksand, and the persistent poison fumes.

But as industry legend has it, von Stroheim scoffed at the handwringing of corporate types and studio naysayers and set out to shoot his film in the Death Valley. And so, he set out into the desert with a caravan of seven trucks and cars carrying his two stars, Gowland and Hersholt, three cinematographers and a violinist (who provided mood music until his instrument cracked from the heat), some three dozen technicians, and just one woman, Eve Bessette, in charge of the script.[15] Far exceeding the proverbial 40 days and 40 nights Christ spent wandering the desert, the crew of *Greed* passed six weeks filming in the dunes and the salt flats, camping on army cots in the open air along the perimeter of Furnace Creek Ranch. The heat set no special records that summer but nonetheless reached an unbearable 52 degrees Celsius in the shade of the cameraman's umbrella out on the salt flats. Von Stroheim rather implausibly claimed the thermostat had reached 120 Fahrenheit (48 Celsius) at times, and Hersholt, crawling over the glaring salt flats, was convinced it had felt as hot as 150 Fahrenheit (65 Celsius).

The ordeal certainly took its toll. Fourteen crewmembers fell ill from the heat and had to be pulled out, while Hersholt, who claimed to have shed 12 kilograms, became delirious with sunstroke and had to be hospitalised for several weeks. It was 'the most terrible experience any of us has ever gone through', Hersholt recalled, and before the filming was finished, there was 'murder … in our hearts'.[16] According to Hollywood legend, toward the end, von Stroheim carried a pistol day and night and forbade anyone else from getting near a loaded gun. By the time they got to the final scene—the fight to the death between Mac and Marcus—both Gowland and Herscholt were befittingly burned, blistered, and bleeding. The seemingly mute and inert desert turned into a hostile entity to overcome—urging producers to rethink other solutions that might ease the production process.

For von Stroheim, the real ordeal began when the filming ended, and he started editing the 200,000 feet of film they had shot over nine months at a cost of $470,000, a princely sum in those days. In January 1924, he emerged from the cutting room with his first version, a 42-reel, nine-hour film, which he planned to show in two parts with a dinner intermission. But before it was released, Goldwyn studios merged with Metro and von Stroheim's nemesis, Louis B. Mayer, took over, demanding a much shorter film. So von Stroheim returned to the cutting room and finally emerged with a version cut to just three hours. But to von Stroheim's dismay, the studio demanded still more cuts.[17]

The hazardous slog and financial strain that the crew and company went through to film *Greed* and to bring it to the silver screen quickly became the stuff of Hollywood legend. As mentioned, the studio bigwigs sought to dissuade von Stroheim from filming in the remote desert out near the border with Nevada and instead advised him to make use of dry terrain nearer to Los Angeles that might simulate the salt planes of Death Valley. Von Stroheim's journey out into the farthest reaches of the Mojave Desert of California clashed with the prevailing perception that there was no need to venture too far from home because, with a little ingenuity, certain locations could be contrived to stand in for a host of places. After all, that was the idea behind the emerging practice of shooting 'on location'.

Indeed, recall that Hollywood itself developed, in part, out of the quest of the established east-coast filmmakers to find new locations that would allow films to incorporate multiple landscapes and places. The emergence of what could be called a cinematic space necessitated locations that could be made to appear as any and every kind of place, generic enough to cater for films depicting scenes in a whole variety of locales. The regulatory, geological and meteoritical potential of the Golden State, alongside the ready availability of wide-open deserts there, paved the way for the development of the film industry in California after 1910 and saw deserts emerge as 'natural sized studios' that could simulate various worlds.

The film director D.W. Griffith and his influential Biograph team led the trend in 1910, fleeing the bitter east-coast winter that was so inhospitable to shooting films on location, to avail themselves of the warm outdoors and abundant winter sunshine in California: "This is when the natural landscape depicted in these post-1910 films became more authentically Western," writes Mary Lea Bandy, "with vast empty deserts places replacing north-eastern lakes and forests."[18] The east-coast filmmakers may also have been on a quest for American locales and backdrops that would give them an edge over their European rivals for novelty and cinematic spectacle. They were looking for a landscape of their own. As Western film expert Edward Buscombe notes:

> [T]he desert of the southwest gave the American Western film a 'unique selling point' as compared to the Western being made in Europe and the Eastern United States by the French Pathe company. Europe too had mountains, but it had no deserts and canyons to rival the spectacular sights of Arizona and Utah.[19]

Making its way west from the colder and darker climate of the east coast, the film industry gradually released itself from the interior spaces of film studios, where filmmakers could control what appeared as backdrop behind the action. Yet initially, the techniques and logic of indoor studio filmmaking, in which backdrops could be changed to simulate various places, were strangely transposed to outdoor desert locations. The film historian Brian Jacobson traces meticulously how deserts continued to be used as a mere backdrop for the action during the early years of Hollywood. Unlike the dynamic, unpredictable, and sometimes hostile environment of real deserts, studios offered filmmakers a highly controlled space to craft their narratives. The lust for locations behind *Greed* urged the film industry to find desert locations nearer and more suitable for long-term and recurring use. While celebrated as a realistic depiction, *Greed* bolstered the perception in the industry that realism should not guide the selection of locations but instead the feasibility of finding look-alikes that could simulate sought-after places. The horror of a desert in charge shocked the studios first into sound stages and later onto landscapes over which they could exercise their power.

The appearance of generalised desert environments in Hollywood films should be considered inextricable to the climate, material conditions and economy catering for the production machine of popular cinema. The desert was adopted as a chameleon geography that could simulate a range of places. Film historian Jennifer Peterson claims that while most film histories explain the move west as the result of the mild climate of southern California and the diverse topography, 'what seemed symptomatic about this list of factors, is the ideology of manifest destiny and settler colonialism that undergirds it'.[20] Indeed, this logic was extended to its cartographic extreme. For example, a 1920s Paramount studio location map sketched all the different places in California that might function as stand-ins for stories set across the world. It indicates, for instance, that the Sacramento River could stand in for the Mississippi River, the canals of the Venice neighbourhood in West Los Angeles could be used to depict the real Venice, and most importantly, the Mojave Desert could be used to represent the Sahara Desert.

An article in the American Cinematographer from 1928 instructs cinematographers trying to cope with the barren desert landscape to take control over the composition by adding props to the frame: 'There's no point in saying, "Well, nature didn't put anything there to complete my composition, so I'll just have to make the best of it" … Your answer is to carry your composition with you'.[21] The solution, borrowed from the logic of studio production, is this:

> If you're in the desert and need a natural looking foreground, you can prop up a cactus or similar desert plant in the place where it will make your composition most effective … make this desert any place with just a touch (figure 4).[22]

Brian Jacobson shows that from its early days, 'cinema depended on nature—in the interaction of sunlight and light-sensitive materials it activated—even as studio architects and engineers sought to disassociate filmmaking from the natural world by selecting, manipulating, and\or simulating its desirable features'.[23] The simulated escape from urban modernity helped attract eastern city dwellers to western film subjects. As Nanna Verhoeff argues, early Westerns succeeded, in part, by spicing up the scenery. Westerns added adventure, offering 'nature as a thrilling and exotic site for escapism for the urban population'.[24]

As audiences warmed to these western settings, film companies sought to capitalise on their popularity by increasing western and southern productions. Exterior locations were more easily accessible due to the increasing mobility afforded by the nation's expanding railway system. As a result, warmer and sunnier locations in California became the new centre of sprouting industries. What was once imagined as the Great American Desert in the west (New Mexico, Arizona and southern California) promised a new gold rush, this time in desert 'locations' there for the taking. Gradually, wariness toward escaping the controlled interior spaces of the studio dissipated, making the Mojave Desert and other parts of California much more attractive substitutes for the painted landscapes within studios.[25]

But the core treatment of the land as a mere setting remained, as it was considered generic enough to serve as a narrative *terra nullius*, a land belonging to no one and therefore up for grabs. 'Filming a place', notes Jacobson, 'often meant owning it too'. Thus, while cinema represented a way to tell a story staged in 'no man's land', production companies often purchased the same land to keep it as a film lot for future productions.[26] Jacobson lingers on a 1914 photograph of Hollywood pioneers Jesse Lasky and Cecil B. DeMille photographed at the newly acquired Lasky ranch in the San Fernando Valley suggests the logic they used to do so. Out on horseback, Lasky and DeMille are, as the caption describes, 'prospecting for locations'.

> The term 'prospect' neatly encapsulates both the emerging idea of 'location'—as a resource that needed to be secured, controlled, and efficiently accessed—and the means by which California filmmakers did so. As it appeared in the caption accompanying Lasky and DeMille's photo, prospecting points to the similarity between prospectors' pursuit of mineral deposits—especially with reference to the California gold rush and Los Angeles's growing oil industry—and filmmakers' search for suitable backdrops.[27]

Indeed, the best of all studios, Jacobson suggests, is nature—and the desert was its ideal. One specific case, which appeared in the weekly silent era film industry magazine *Moving Picture World* in 1911, warned young filmmakers about locations becoming so frequently reused that even audiences were increasingly noticing backdrops across different films, proclaiming that 'we exhibitors become as familiar with film makers studios as with our own backyards', and 'feel the repetition of studio settings'.[28]

Places that could become other places were considered especially useful. Scenes in the desert offered aesthetic alternatives to painted sets and provided an experience of nature for urban dwellers, those 'city shut-ins' who remained embedded within the artificial environments of the modern metropolis. The desert opened up a playground for the industry to make use of its vast spaces as one of the few geographies that can still function as a stage for the frontier narratives that once inspired Hollywood producers to venture into southern California, Arizona and Mexico.

Duel in Arrakis

Exactly 100 years after von Stroheim shot his duel in Death Valley, an equally intense (and even more expensive) film production moved into the desert to capture yet another duel. Washed in a yellow tint and with an unforgettable and formidably evocative soundtrack composed by Hans Zimmer, Denis Villeneuve's *Dune: Part One* (2021) brings to the silver screen the desert world of Arrakis, adapted from the science-fiction novel by Frank Herbert. Arrakis, an arid and desolate planet, is the only source of the most valuable substance in the known universe, the spice melange. The protagonist, Paul Atreides, played by Timothée Chalamet, is at once heir to one of the great noble houses that seek to exploit the spice and a messiah fated in ancient prophecy to lead the planet's indigenous inhabitants, the Fremen, to liberate Arrakis. But to do so, he must learn to become part of the desert and, in the words of his father, Duke Leto Atreides (Oscar Isaac), 'harness desert power'. Infused with a righteous white saviour pathos and spiced with orientalist depictions of the Native people, *Dune* essentially recasts the colonial perception of the desert as an ex-territory, a realm of radical individualism and savage conflict over scarce resources.

Towards the film's end, Paul finds himself amidst the Fremen, some of whom suspect he is the Mahdi (or messiah). Jamis (Babs Olusanmokun), a formidable Fremen warrior, disputes Paul's membership in the group and challenges him to a ritual duel to the death. The challenge is accepted, and the fight begins (Figure 1.4). Like Goya's fighters, clubs raised, ready to strike the mortal blow, Paul and Jamis face-off, eyeing each other warily, brandishing blades rather than wooden cudgels. While the weapons differ, the mortal drama is the same. Importantly, Paul must seize this chance to earn his place among the Fremen and fulfil his destiny as the Mahdi. Getting the better of Janis but hesitant to take his life, as he must, Paul hears a voice whisper to him, as if from the desert itself, calling him to channel his power

FIGURE 1.4 Still from *Dune* (Denis Villeneuve, 2021). Alamy Stock Photo.

to meet his destiny. The camera pans across Paul's face as he resolves to 'harness desert power', strike the mortal blow and kill his opponent. Perhaps not unlike *Greed*, the desert is once again a hostile force to reckon with; a backdrop imbued with a deadly agency.

To capture the unworldly quality of Arrakis, an arid terrain fraught with peril, the production company Paramount Pictures ventured far beyond the deserts of California to Wadi Rum in Jordan. The geological features and the topography of Wadi Rum, together with the high concentration of iron oxide, which causes the desert surface to take on a reddish colouration, make the location a particularly suitable backdrop for films depicting fictive dystopian planets. But the economic incentives provided by the Jordanian government and its recognition by UNESCO as a World Heritage Site make it most attractive to Hollywood studios. In 2003,[29] Jordan's government established the Royal Film Commission, under the aegis of Prince Ali bin Hussein, a senior member of Jordan's royal house, to advance cinema production in the country and promote it internationally. According to its former director George David, Wadi Rum has long been a destination for high-budget Hollywood cinema, dating back to David Lean's 1962 film *Lawrence of Arabia*, one of the films that really put cinematic deserts to the fore.[30] To bring in high-budget film productions, the scheme originally required a minimum expenditure of $1 million in Jordan. The maximum rebate granted to a foreign production has since been lifted to $2 million and must be claimed by a production service company registered in Jordan.

Since its launch, the Jordanian government has raised the upper limit of the cash rebate on eligible productions filming in Wadi Rum to 25 percent of all qualified expenses on film, television, and other media productions incurred within Jordan. To qualify, a film production company must employ at least 50 Jordanian crew members and train 20 interns. The rebate is in addition to sales tax exemptions for products and services purchased within Jordan, customs taxes and duties exemption for equipment and material imported into the kingdom, and exemptions on salaries and fees for non-Jordanian cast and crew members. In 2015, Prince Ali bin Hussein claimed that 'the film industry has created more than 8,000 jobs in the past seven years and, from 2007 until today, almost $140 million was spent in Jordan.'[31] Effectively, this financial scheme has created a demarcated zone that operates as a tax haven and caters mostly to wealthy American producers. The nature-sized 'studio' of Wadi Rum is an asset for the Hollywood film industry, which has no problem meeting the prescribed spending minima.

Significantly, in 1985 UNESCO recognised the city of Petra and large parts of Wadi Rum as a World Heritage Site, effectively banning extractive industries from drilling, digging, or mining in the desert. In 1996, following the peace agreement with neighbouring Israel, the World Bank proposed what it called the 'Second Tourism Development Project' to spur investment in environmentally sustainable tourism in Jordan. The plan aimed to bring tourism management and tourism-related activities under one umbrella instead of the disaggregated management approach by local Bedouin groups while underlining new environmental protocols. The Wadi Rum Protected Area (WRPA) has the status of a protected archaeological site under Jordan's Law of Antiquities. It is administered as part of the Aqaba Special Economic Zone Authority, a quasi-autonomous body responsible for the sustainable management and economic development of the entire region. The use of the desert by Hollywood is thus an exemption to an entire regime restricting industrial activity in the area in the interests of long-term sustainability.[32]

This 'Hollywood exceptionalism' in the Jordanian desert once again calls attention to Serres' injunction that we ignore at our peril the material surroundings in which the striving of humanity takes place. The quicksand referenced from Goya's painting is recast in this context as the material circumstances that facilitate film production. The quicksand is the location itself, together with the geographical and geopolitical properties supporting the staged scene. Moving downwards through the fictional world of Arrakis, we crash into the sand of a very real desert. Moving downwards—or 'looking under the table', as Kracauer put it—means departing from the represented desert and probing into the material conditions that made Wadi Rum a good stand-in for Arrakis in the first place.

Wadi Rum is only one example. Morocco, Israel and the UAE have also established diverse legal and economic configurations to make it worthwhile for Hollywood to shoot in their deserts. Throughout the 1980s, Israel–Palestine was a favoured site for shooting action war films such as *Rambo III*, *Delta Force* and *Iron Eagle*. Hollywood producers were delighted when the Israeli Defense Forces (IDF) rolled out the red carpet for the studios to set up in deserts appropriated by the IDF for combat training. They made good use of the militarised infrastructure in these zones as props when filming depictions of American 'good guy' commandos facing off against all manner of Arab 'bad guys', be they states or terrorist organisations in Iraq, Afghanistan and other Arab countries. The unlikely participation of the IDF in the production of the films inaugurated a unique military-industrial environmental configuration, combining the ecological properties of the region, Israel's hold over certain desert territories and the country's diplomatic relations with the United States.

As I have suggested elsewhere, this combination grew out of the colonial gaze that, throughout the twentieth century, framed the Negev Desert as neglected and barren land to be redeemed, cultivated and settled.[33] For decades Israel has continued to sustain the cultural image of this desert as uninhabited land, abandoned and inhospitable, a perception that has assisted the IDF in gradually seizing expansive desert territories and declaring them as closed military zones. This rapid militarisation of the desert and the cultural imaginings that have justified it have eased the way for Hollywood film companies to produce big-budget narrative cinema. Legal provisions enacted by Israel for this specific purpose make the desert available for the film industry as a kind of narrative laboratory, where tales of human survival in extremely hostile conditions are readily visualised. Such desert locations are used to test new technologies of visualisation and effects that are otherwise difficult to produce. Thus, cinematic technologies, securitised desert landscapes and neo-colonial imaginaries congeal on the desert surface to divest it from the human and nonhuman inhabitants that comprise its native ecology.

In Morocco's city of Ouarzazate in the Sahara, a combination of cheap labour and vast spaces convinced a local entrepreneur in 1983 to set up Atlas Studios, the largest film lot in the world by surface area, which includes giant props from several Hollywood blockbusters, from *The Jewel of the Nile* (1985) to *The Mummy* (1999). Hollywood cinema, vast open desert, and global tourism cluster around Ouarzazate, dubbed 'the door to the desert,' lending Atlas Studios a certain orientalist sheen as a kind of 'desert Disneyland' of cinematic spectacle. Recently, films such as *Transformers* (2007) and the series *Game of Thrones* (2011–2019) have exploited a terrain that can be readily abstracted to bring unworldly imaginaries to the silver screen. Similarly, the UAE has recently attempted to bring Hollywood to its deserts with attractive free trade zones and tax rebates designed for high-budget Hollywood cinema.

Connecting the various titles are desert environments adopted as a stage for fiction; an arid land, historically relegated as a wasteland, converted into Euro-American testing zone for spectacles of war and doomsday. In its attempt to preserve the contrived fantasy of an uninhabited frontier, big-budget cinema relies on territories closed off for techno-military experiments, dispossessed from Indigenous people, and made into deregulated economic zones and tax-free areas.

Conclusion

The desert is a laboratory for frontier narratives. No doubt, Hollywood cinema continues to demonstrate an intense preoccupation with visualising the notion of the frontier. Yet at the heart of such frontiers lies a paradox that defies representation; the frontier, after all, was never a place but a flexible ideological, economic and cultural concept fluctuating with the rapidly expanding United States. In his re-examination of Frederik Jackson Turner's invocation of the 'frontier' concept, historian Greg Grandin notes that Turner's usage radically transformed the meaning of the word itself.[34] With the movement of white settlers westwards, the concept of 'frontier' was finally released from its literal meaning of a border or a threshold to take on a more abstract sense. 'Frontier became fuzzier,' writes Grandin, eventually indicating 'a cultural zone or a civilizational struggle, a way of life: a semantic change electrified by the terror and bloodshed that went along with settler expansion.'[35] Turner himself wrote in 1893 that the 'term is an elastic one [and] for our purpose does not need sharp definition.' He then went on to define 'frontier' in at least 13 different ways to indicate, among other things, 'a return to primitive conditions,' 'the outer edge of a wave,' 'the meeting point between savagery and civilisation,' something that 'lies at the hither edge of free land', a 'harsh environment' and a 'field of opportunity.'[36] Is the desert not a 'field of opportunity' for the film industry, where this pioneering myth can be best visualised?

In an overpopulated world, in which borders are fixed, the frontier narrative looks outwards, where financial speculation and resources can still be imagined as untouched. In other words, Hollywood needs new deserts, and those deserts must be foreign to the American eye if they are to depict new frontiers. This unstable meaning of 'frontier' already implicates the desert landscape. Rather than as a backdrop or referential place, the landscape becomes an indicator of the invisible line that separates 'wild' from 'civilised'. Moreover, it refers to, or points to, the fact that this line is moveable, that the conquest at its heart is precisely the progression of a borderland that will eventually dissolve when all is conquered. Nanna Verhoeff has suggested that the resulting desert landscape in early Hollywood cinema, 'whether principal or subordinate, beautiful or barely visible, is the visible evidence not of a Western landscape but of a fiction, a cultural projection, a fantasy called "West"'.[37] This fantasy is particularly appealing in a time of rapid change. Modernity, with its shocking transformations of every aspect of social life, was particularly receptive to the sense of space this contrived 'West' and its primary embodiment, landscape, emanated for delectation, thrill and reassurance alike. Today this 'contrived West' is projected upwards, out to the planetary scale. Done with Earth, the border expands and with it, the desert. Storied desert could thus be understood as the spaces of speculation, where the founding myth of the West can find room to persist and expand.

To understand cinema's reliance on deserts, we must not remain within the limits of the cinematic frame to ask what the cultural significance of the representation of the desert is. Instead, we should dig deeper to expose what lies beyond the frame's limits to reach the material relations that cinema retains with desert spaces. I have suggested in this chapter that the assemblage of laws, environmental imaginaries and economic incentives continue to demarcate the desert as a narrative laboratory, locked between the spectacularly visible and the long durée of environmental transformation. Moreover, high-budget cinema continues to construe the desert as an exteriority, where the industry can represent the colonial dream of a tabula rasa while exploiting neo-colonial economic incentives in deserts worldwide.

Abstracted and defused, the fictionalised lands in which epic duels take place are a manifestation of deserts everywhere, perceived at the edge of empire, at once dead yet teeming with life that should be preserved. Today we must revisit desert geographies that have been, and continue to be, obfuscated through the spectacle of empire. Now that the desert is no longer easily tucked away by European imaginaries and slowly demands its place in our material worlds, there is an urgency to unlearn the fallacious coupling of deserts and cinematic spectacle to reposition the arid land within our common imaginary of Earth.

Notes

1 Rory Carroll, "Hollywood and the Downwinders Still Grapple with Nuclear Fallout", *The Guardian*, 6 June, 2015, https://www.theguardian.com/film/2015/jun/06/downwinders-nuclear-fallout-hollywood-john-wayne
2 Terrence R. Fehner and Francis George Gosling, *Origins of the Nevada Test Site*. History Division Executive Secretariat Management (US Department of Energy, 2000).
3 Ibid. 20.
4 Serenella Iovino and Serpil Oppermann, "Theorizing Material Ecocriticism: A Diptych." *Interdisciplinary Studies in Literature and Environment,* 19 (October 2012): 448–475.
5 Michel Serres, *The Natural Contract* (The University of Michigan Press, 1990), 2.
6 Ibid. 3.
7 Ibid.
8 Siegfried Kracauer, *Theory of Film: The Redemption of the Physical Reality* (Oxford Press, 1960), 42.
9 Ibid. 43.
10 Elana Gorfinkel and John David Rhodes, *Taking Place: Location and Moving Image* (Minnesota University Press, 2011), 5.
11 Ibid. 6.
12 Serres, *The Natural Contract*, 4.
13 The tension between those terms, space and place, has been theorised by Michel de Certeau. Place, for de Certeau, is "an instantaneous configuration of positions"; or, "an indication of stability." Place is the antithesis of movement and therefore of change and (revolutionary) possibility. Space names, for de Certeau, the medium through which such change is possible. Space is "actuated by the ensemble of movements deployed within it." De Certeau argues that a place is like "the street geometrically defined by urban planning" that is "transformed into a space by walkers." Often quoted is anthropologist Marc Augé's conception of the nonplace, "a space which cannot be defined as relational, or historical, or concerned with identity." But Augé's conceptualisation of the nonplace possesses a potentially colonising and ontological nonspecificity not unlike the nonplace itself. My intension here is to resist this nonplaceness that cinema creates. Perhaps closest to this conception of cinematic space is Kracauer, for whom a cinematic place "opens up a dimension much wider than that of the plot". The details for Kracauer are entry point to material places, that can even interfere with the narrative. See de Certeau, Michel. *The Practices of the Everyday* (Third Edition) (UC Press, 2011) (organically published 1980).
14 Jonathan Rosenbaum, *Greed*, BFI Film Classics series (Palgrave Macmillan, 1994).
15 Ibid. 10.
16 Ibid. 15.

17 Richard Lingenfelter, *Death Valley & the Amargosa: A Land of Illusion* (University of California Press, 1986), 65.
18 Mary Lea Bandy and Kevin Stoehr, *Ride, Boldly Ride: The Evolution of the American Western* (UC Press, 2012), 16.
19 Edward Buscombe, *Unforgiven* (BFI Classics, 2004), 45.
20 Jennifer Peterson, "The Silent Screen 1895–1927," in *Hollywood on Location: An Industry History*, eds. Joshua Gleich and Lawrence Webb (Rutgers University Press, 2019), 23.
21 Jerome Ash, "Carry Your Compositions with You," *American Cinematographer* (1928), https://archive.org/details/americancinematographer?&sort=-date&page=12
22 Ibid.
23 Brian R. Jacobson, *Studios Before the System: Architecture, Technology, and the Emergence of Cinematic Space* (Columbia University Press, 2015), 16.
24 Nanna Verhoeff, *The West in Early Cinema: After the Beginning* (Amsterdam University Press, 2006), 48.
25 Jacobson, 174.
26 Ibid. 175.
27 Ibid. 187.
28 Ibid. 185.
29 Official site of the Jordanian Government, the Hashemite Kingdom of Jordan, Department of Antiquities, https://portal.jordan.gov.jo.
30 Raied Shuqum, "Jordan Hopes to Become Major Filming Destination," *Arab Weekly*, July 24, 2015, https://thearabweekly.com/jordan-hopes-become-major-filming-destination.
31 Ibid.
32 See the expanded article on this particular use of location in Jordan in the article Daniel Mann, "Red Planets: Cinema, Desert and Extraction," *Afterimage* (UC Press, 2021), https://online.ucpress.edu/afterimage/article/49/1/88/120349/Red-PlanetsCinema-Deserts-and-Extraction
33 Daniel Mann, "Rambo in the Dead Lands: On Cinema, War and Climate." *Screen*, 61 (3) (Autumn 2020): 368–384.
34 Greg Grandin, *End of a Myth: From the Frontier to the Border Wall in the Mind of America* (Metropolitan Books, 2019).
35 Ibid. 116.
36 Ibid.
37 Verhoeff, The West in Early Cinema, 205.

Bibliography

Ash, Jerome. "Carry Your Compositions with You." *American Cinematographer*. Autumn 1928.

Bandy, Mary Lea and Stoehr, Kevin. *Ride, Boldly Ride: The Evolution of the American Western*. UC Press, 2012.

Buscombe, Edward. *Unforgiven*. BFI Classics, 2004.

Carroll, Roy. "Hollywood and the Downwinders Still Grapple with Nuclear Fallout." *The Guardian*, 6 June, 2015. https://www.theguardian.com/film/2015/jun/06/downwinders-nuclear-fallout-hollywood-john-wayne

de Certeau, Michel. *The Practices of the Everyday* (Third Edition). UC Press, 2011 (organically published 1980).

Fehner, Terrence R. and Gosling, Francis George. *Origins of the Nevada Test Site*. History Division Executive Secretariat Management. US Department of Energy, 2000.

Gorfinkel, Elena and Rhodes, John David. *Taking Place: Location and Moving Image*. Minnesota University Press, 2011.
Grandin, Greg. *End of a Myth: From the Frontier to the Border Wall in the Mind of America*. Metropolitan Books, 2019.
Iovino, Serenella and Oppermann, Serpil. "Theorizing Material Ecocriticism: A Diptych." *Interdisciplinary Studies in Literature and Environment*, 19, (October 2012): 448–475. https://academic.oup.com/isle/article-abstract/19/3/448/687320?redirectedFrom=fulltext
Jacobson, Brian R. *Studios Before the System: Architecture, Technology, and the Emergence of Cinematic Space*. Columbia University Press, 2015.
Kracauer, Siegfried. *Theory of Film: The Redemption of the Physical Reality*. Oxford Press, 1960.
Lingenfelter, Richard, *Death Valley & the Amargosa: A Land of Illusion*. University of California Press, 1986.
Mann, Daniel. "Rambo in the Dead Lands: On Cinema, War and Climate." *Screen*, 61 (3), (Autumn 2020): 368–384.
Mann, Daniel. "Red Planets: Cinema, Desert and Extraction." *Afterimage*, 49 (1), (2022): 88–109.
Peterson, Jennifer. "The Silent Screen 1895–1927." In *Hollywood on Location: An Industry History*, eds. Joshua Gleich and Lawrence Webb. Rutgers University Press, 2019.
Rosenbaum. *Greed*. BFI Film Classics Series. Palgrave Macmillan, 1994.
Serres, Michel. *The Natural Contract*. The University of Michigan Press, 1990.
Shuqum, Raied. "Jordan Hopes to Become Major Filming Destination." *Arab Weekly*, July 24, 2015. https://thearabweekly.com/jordan-hopes-become-major-filming-destination.
Verhoeff, Nanna. *The West in Early Cinema: After the Beginning*. Amsterdam University Press, 2006.

2

THE DARK, DEAD CORNERS OF THE EARTH

The Imaginary of the Antarctic as "Deserta"

Rune Graulund

While the word desert routinely conjures images of heat and sand, camels and dusty canyons, technically speaking the Antarctic is the largest desert on the planet. Moreover, as the only continent entirely uninhabited by humans and with an average annual precipitation level well below the 250 mm geographers tend to use as the demarcation point for desert regions, the Antarctic lives up to the definition of being a 'desert' in more ways than one. Largely considered to be non-profitable due to lack of easily accessible extractive resources, with minimal precipitation and presenting an environment so harsh that it cannot in and of itself sustain human life over longer periods of time, the Antarctic can in a sense be said to encapsulate the very essence of the Latin roots of 'deserta'. It is deserted, abandoned, forsaken and defined by negatives and all that which it is not. Situated at the supposed end of the world, as far away from the dominance of the Global North as one can be while still remaining on the planet, the Antarctic can be said to be not just the largest, but possibly also the most desert of deserts.

As a consequence of its remoteness, emptiness and harshness, imperialist and capitalist interest in the Antarctic was negligible until the late nineteenth century, to the point of being non-existent. The spectacular race for the South Pole at the beginning of the century aside, any consistent attempt at establishing imperialist and national claims on the Antarctic would not be made in earnest until the postwar period, resulting in the first Antarctic Treaty in 1961. With the US, the USSR, Britain, Belgium, Japan, France, Chile, Norway, Argentine, Australia, New Zealand and South Africa recently taking an interest in the continent, the sudden scramble to claim territory on the Antarctic was however triggered more by the fear of loss of prestige, or an as yet undiscovered future potential of the region rather than by any actual and immediate plan for profit and extraction. To this day, while global warming has begun to free parts of the Arctic from centuries if not millennia of ice,

DOI: 10.4324/9781003397212-4

evincing natural resources such as oil and minerals in the most northern extremes of the hemisphere, which in turn has led to yet another scramble for dominance among the Arctic nations,[1] the Antarctic has as yet to experience any comparable rush to monetize its resources.

Unexplored and unvisited by humans until very recently, the Antarctic has however acted, and continues to act, as a particularly rich space for the imagination. It is through visions of the Antarctic as a space empty of the human but rich with the alien, the monstrous and the non-human that I will in this chapter read the Antarctic in the tradition of the '*terra incognita*' that cold as well as warm deserts were traditionally considered to be. As European explorers and settlers encountered arid landscapes around the world, whether it was in sweltering Africa, Asia, Australia or the Americas, or in the freezing, impenetrable 'wastes' of the Arctic and the Antarctic, the apparent lack of life would typically lead to the conclusion that arid landscapes were either of minimal importance, or even a hindrance, to imperialist, colonial, extractive and capitalist projects. With the discovery and increasing need for oil, rare earth minerals, uranium and the like, combined with the refinement of irrigation technologies that would turn formerly infertile desert lands in for instance the American Southwest into farmland, the position on the unprofitability of arid environments would eventually shift. Of all the many and varying environments encountered by the subjugation of the ravenous forces of colonialism, imperialism and of capitalism, arid landscapes would typically be last in line to be subjugated, simply because other, more easily exploitable environments were considered ripe for taking. While arid landscapes have often been perceived to be terrible, or 'simply boring',[2] they have for much the same reason been considered a space in which many things otherwise out of the ordinary could be imagined to take place. Where little actual experience is present, the imagination has a tendency to step in and take over.

This chapter will start out with a general explorative and intellectual history of the Antarctic as it has been documented and imagined through the centuries. In this, it will be contrasted at length with the Arctic which, though at first sight seemingly similar as one of two polar regions, in fact has a very different conceptual as well as colonial history. From this comparative reading of the Antarctic as the negative ('anti') or opposite of the Arctic, this chapter will continue to read a range of texts and films in which the Antarctic has been envisioned as an otherworldly space, one of the 'earth's dark, dead corners' that it 'is absolutely necessary [is] left alone',[3] but also a space that has for those same reasons continued to fascinate explorers real as well as imagined. In this, this chapter will argue, the Antarctic is read as so many other deserts of the world, which is to say as a blank canvas ready for inscription, but also as a mirror, a passage to other worlds and realities, and to other states of being.

The second main concern of this chapter is the overarching tragedy of the Antarctic, which is that it is disappearing, or at least rapidly changing form. Like its northern counterpart, transforming and disappearing at an even faster and more alarming pace, the Antarctic is heating up, its ice and snow disappearing. As I will argue throughout this chapter, both regions have thus become the canaries in the coalmine of planetary emergencies; or if one will, the penguins and the polar bears of a fossil-fuelled mine of the present that is threatening to blow up the future. Starting with the ozone hole layer scare in the 1980s, the polar regions have continued to represent the vulnerability of the planet, which is deeply ironic considering their traditional status as the toughest, most dangerous environments on Earth. With the most recognizable and reproduced visualization of the climate emergency likely being that of a polar bear on a melting ice floe, the Arctic has been the most heavily publicized of environments under threat from a rise in global temperatures, but the Antarctic is close on its heels. As such, the totem animals of this chapter, and possibly of this planet, may be the unlikely pairing of the penguin and the polar bear; a weird but ultimately telling parable of a world that initially got into trouble by bringing canaries into coal mines in order to exploit the very chthonic reserves of carbon that would kick off the slide towards the fossil-fuelled atmospheric nightmare that is our current predicament.

Liminal: Arctic vs. Antarctic

It is imperative to stress that the account above is the perspective on arid environments as it was experienced from a *Eurocentric perspective.* The Indigenous peoples of the American or Australian deserts, of the Rub' al Khali, the Gobi, the Kalahari or the Sahara of course did not view the lands they considered as home to be either empty, terrible, or unfathomable, or indeed as necessarily specifically fantastic. This is however precisely also why the Antarctic stands out in a roster of geographical locations that might be termed 'desert'; not just to its etymological other, the Arctic, to which the Antarctic is in its very essence defined as being 'in opposition' (Ant-Arctic), but to all arid environments, indeed to the world as such. 'From the beginning, it had been crucial to its imaginative impact as a place that it was a place apart, locked away at the bottom of the world,'[4] Francis Spufford remarks in the introduction to *The Antarctic* (2007), an edited collection of texts testifying to explorative engagements with the continent from the late nineteenth century and onwards. While modern technology has since then made it considerably less deadly and challenging, the Antarctic is nevertheless still inhospitable enough that the 'perception of the Antarctic as "the last great wilderness" is, today, commonplace'.[5]

As an environment harsh enough that it was for a long time capable of holding modernity at bay, it is also important to point out that the Antarctic is the only continent in the world with no Indigenous or indeed any permanent human population. The Antarctic can therefore be said to stand as 'deserta ultima' in comparison both to its icy counterpart in the north, and to the other, warmer deserts of the planet. Most hot deserts of the world are inhabited to some extent or other by nomadic cultures, as is the Arctic, where Indigenous Inuit people, repressed though they often were (and are) by colonizing forces, can nevertheless counter the speculations of both temporary visitors and more permanent settlers with a deep cultural tradition of what life in the Arctic is actually like as day-to-day experience rather than an exercise in imaginaries. The Antarctic stands in stark opposition to this in that no human habitation has succeeded to survive on the continent for longer periods of time without resources, sustenance and fuel imported from outside. Accordingly, no Indigenous cosmologies exist to challenge the visions of the Eurocentric explorers, armchair travellers and fictional accounts of the region.

Meager as it may be in terms of agrarian potential, mineral wealth and Indigenous storytelling, the Antarctic is for those very same reasons also uniquely positioned in the Eurocentric cultural tradition as a space that is *terra incognita*, *terra nullius* and *tabula rasa*. Unknown and uninhabited, void and unfathomable, cold and (supposedly) dead, the Antarctic stands as a space where nothing and everything is imagined to happen, and where the boundary between this world and the next breaks down. From horror classics like H.P. Lovecraft's *At the Mountains of Madness* (1936), on to John Carpenter's *The Thing* (1982), up to and including one of the latest films in the Godzilla franchise, Adam Wingard's *Godzilla vs Kong* (2021), the Antarctic has been imagined as a space outside of history, as a 'land that time forgot', but also often as a gateway to other worlds, either subterranean (Lovecraft, Wingard), extraterrestrial (Lovecraft, Carpenter), or ghostly, 'a fundamentally unhomely place'[6] in which the dead and the dying feature more prominently than the living. It is telling that in the aftermath to the so-called 'race to the South Pole,' posterity has cherished not so much the successful Norwegian explorer Roald Amundsen who *did* in fact in 1911 make it to the South Pole (and back), but to British Robert Falcon Scott, who perished on the ice alongside the other members of his doomed expedition.

Contrast this again to the Arctic. Partly due to Canadian, English, Danish, Norwegian, Russian and (lately) American imperial interests in the region, and partly due to an Indigenous presence stretching back since time immemorial, the Arctic has been far more thoroughly mapped, explored, experienced and therefore also *storied* than the Antarctic. 'Though the Arctic comprises some of the most inhospitable terrain on earth, people have been living there for thousands of years.'[7] As English explorer and travel writer Sara Wheeler points out, contrasted to the Antarctic, 'a metaphor for a *terra incognita*', the Arctic is if not exactly home, then at least much closer to home, literally as well as figuratively speaking. 'The Arctic is our neighbour; part European; part North American; us,' she says, hence in a comparison to the Antarctic, which 'existed most vividly in the mind', the Arctic is 'complicated, life-infested.'[8] Accordingly, 'the literature of the Arctic is immense,'[9] storied and historied, experienced and recounted, colonized and visited.

The Arctic has, however, like the Antarctic, traditionally been considered a liminal rather than a familiar space, 'a space that can be viewed, for the majority of people, only from the outside'.[10] Markedly different than the Antarctic in that it has a long history of an actual rather than an imagined 'contact zone'[11] between explorer and Indigenous populace, wildlife and land, the Arctic has as often as not still been considered a space in which reason and knowledge ends and madness and the mysterious begins. 'Direction is suspended at the North and South Poles; they are places outside place',[12] Peter Davidson remarks in *The Idea of North* (2005). As Roland Huntford points out, the Arctic was 'as hidden as securely as the dark side of the Moon [and therefore] a place of speculation'[13] all up until the twentieth century. This sentiment of the directionless and speculation of the polar regions is expertly satirized by the 'Tundra' episode of British comedy duo *The Mighty Boosh*. Introducing their 'Tundra Rap', the characters Howard Moon and Vince Noir discuss the mysterious draw of the icy ends of the world:

HOWARD MOON: You know what it is about this place that gets to a man?
VINCE NOIR: Not really?
HOWARD MOON: Have a look through here [hands Noir binoculars]. What do you see?
VINCE NOIR: Nothing?
HOWARD MOON: Exactly. It's the nothingness. The whiteness. The endless…ness. Stretching on beyond the human imagination. Desolation of the soul.

(SE1E04 00:10:42–00:11.02)

The draw of nothingness and lack of purchase with the human imagination is also in play with the Norwegian explorer Fridtjof Nansen (1860–1930), albeit combined with a strong and somewhat paradoxical urge to claim intimacy with it. 'Our ancestors, the old Vikings, were the first Arctic explorers,'[14] Nansen claims, making a personal, regional and national claim on the Arctic that goes far back in history. 'Long before other seafaring nations had ever ventured to do more than hug the coast lines, our ancestors had traversed the open seas in all directions, had discovered Iceland and Greenland, and had colonized them,'[15] Nansen boasts. In this, Nansen is positioning himself and his (Scandinavian) ancestors above and beyond those with no such pedigree, no such knowledge of the north and no such 'national character.'[16] Yet for all the supposed familiarity and history of 'our' long and intimate history with the Arctic, Nansen ultimately also cannot but fall back on the familiar trope of the Arctic as a place outside of history and of knowledge, claiming the Arctic to be at one and the same time familiar and strange, known and unknowable:

> Unseen and untrodden under their spotless mantle of ice the rigid polar regions slept the profound sleep of death the from the earliest dawn of time. Wrapped in his white shroud, the mighty giant stretched his clammy ice-limbs abroad, and dreamed his age-long dreams.
>
> Ages passed – deep was the silence.
>
> Then, in the dawn of history, far away in the south, the awakening spirit of man reared its head on high and gazed over the earth. To the south it encountered

> warmth, to the north, cold; and behind the boundaries of the unknown it placed in imagination the twin kingdoms of consuming heat and of deadly cold.
>
> But the limits of the unknown had to recede step by step before the ever-increasing yearning after light and knowledge of the human mind, till they made a stand in the north at the threshold of Nature's great Ice Temple.[17]

Here, Nansen is positioning not just himself and his fellow Norwegians as inheritors of 'the awakening spirit,' but of the European tradition as such. Writing in 1897, Nansen is like so many if not indeed all of his (European) contemporaries framing the Arctic in an imperial discourse of domination, a stage on which the national mettle and character of explorers, conquerors and colonists can duke it out in what he perceives to be an empty arena free of the confusing noise of civilization further south. That is to say as a *tabula rasa*, a '*great unknown region*,'[18] to 'unknown lands, over an open, rolling sea, where no ship, no boat had been before',[19] clear of interference, but also posing a formidable obstacle and a threat, hence as a stage where one (or rather he) could establish a civilizational hierarchy of the bravest, the toughest and the most intrepid. Tellingly, Nansen quite literally frames this quest for imperial glory in the liminal discourse of deserts: of the 'consuming heat' of the deserts of the south and the 'deadly cold' of the great Arctic desert of the north.

Transgressive Territories: Subterranean/Extraterrestrial, Alien/ Albatross

The liminal character of the polar regions is richly documented in fiction. In Mary Shelley's *Frankenstein* (1818), the majority of the plot unfolds on the European continent. Yet the overarching tragic story of Doctor Frankenstein and the creation, abandonment and subsequent battle with his 'creature' is bookended by a frame narrative that is set in the Arctic. An epistolary novel addressed to 'Mrs. Saville', the sister of captain Robert Walton, who at the beginning of the book sets off for 'the sight of a part of the world never before visited', Walton's purpose in going north is rather vague apart from the desired admiration of his sibling. He does so partly in order to sate his 'curiosity', partly to gain glory in discovering a passage through the ice, or perhaps 'ascertaining the secret of the magnet'.[20] A self-proclaimed romantic, Walton achieves none of these things, but he does meet Doctor Frankenstein, 'a noble creature' and a 'glorious creature,' but also his nemesis, 'a dæmon,' and a 'fiend' that has 'the stature of a man, but apparently of gigantic stature.'[21] That it is in 'these wild and mysterious regions' that the final showdown between what Walton perceives as the outer limits of humanity is of course no coincidence, for at the 'most northern extremity of the globe' where the 'cold is excessive,'[22] everything in fact seems to be, precisely, in excess. Overpowering and overflowing, without limit and boundary, it is only natural that Doctor Frankenstein's experiments in the unnatural will come to the fore in a landscape that is at one and the same time considered to be the wildest of the wild, the last and only authentic of landscapes still to give in to the relentless pressures of modernity, but also as a place that is mythical, transcendent and otherworldly.

Two centuries after the publication of *Frankenstein* (2018), the AMC television series *The Terror* (2018) presents the Arctic in more or less the same light. Inspired by the historical expedition of Sir John Franklin's 1845–1848 unsuccessful attempt to locate the northwest passage in the ships the HMS *Erebus* and the eponymous HMS *Terror* from which the series takes its name (and tone), the series portrays the slow but steady decline towards death and madness that would result in the death of every single member of the 129 strong crew of the two ships. The eerie polar setting, the unforgiving cold and a lack of clear visibility are only exacerbated by the cramped quarters of the two ships where the crew find themselves trapped for years on end in the ice, both ships icebound. As if the historical facts were not horrible enough—with the crew and officers growing increasingly hungry and desperate to the point where they resolved to cannibalism—a mysterious creature half polar bear, half demon, also materializes, finishing off the men who are not already figuratively as well as literally at each other's throats. In *Frankenstein* as in *The Terror*, the polar landscape is employed both as a means and a setting in which monstrosity within as without finds its true expression. Just as Frankenstein's limitless ambition leads to his abject failure as a father to take care of the 'creature' he has cobbled together from the bones, meat and skins of a plethora of corpses, a monstrous act of naked ambition that results in a material 'monster' that is as pitiful as it is terrible, the steely ambition of the leadership of the two ships in *The Terror* leads them to face monsters that are at ones and the same time material and abstract, banal and otherworldly.

'I shall kill no albatross,'[23] Shelley's Walton writes to his sister as he journeys northwards, referencing what is likely the most famous poem about the Antarctic in the Western canon, Samuel Taylor Coleridge's 'The Rime of the Ancient Mariner' (1798). Originally published two decades prior to *Frankenstein*, Coleridge's poem recounts the story of a ship driven towards Antarctic waters by a storm. The ancient mariner of the tale draws the ire of the crew when he for no apparent reason decides to shoot an albatross that seems to be leading the ship out of its predicament in the icy waters. What follows is a strange and doomed narrative full of uncanny and grotesque encounters, all of it seemingly provoked by the mariner's inexplicable initial transgression of killing hapless wildlife, leading them with no guide out of 'the slimy sea' of the southern reaches. In his attempt to reassure his sister that he shall perform no such random acts of violence, Walton is thus evoking the liminal nature of the polar regions not only as a space in which the rational and the civilized character of people are tested to their utmost but also as a territory in which the passage of transgression is embedded in the landscape itself, a 'slimy' space in which borders are easily transgressed, and difficult to uphold and police.

This is physically embodied in a range of narratives in which the polar regions are imagined as gateways to other worlds. Jules Verne's *Journey to the Center of the Earth* (1864) describes the passage of Professor Lidenbrock and his companions first to Iceland and then to and through a volcano that leads them deep into the Earth where they discover an underground world of dinosaurs and an underground sea. Similarly, in Peter Høeg's novel *Miss Smilla's Feeling for Snow* (1992), we are presented with an Arctic where communion with the otherworldly happens more easily than in the populous regions further south, albeit in this instance it is extraterrestrial rather than subterranean. The novel begins in Copenhagen but eventually migrates north, to an island off the coast of Greenland, where the protagonist Smilla unravels a conspiracy centred around a mysterious meteorite embedded in a glacier. The concluding chapters of the novel uncover a wild conspiracy of the excavation of the meteorite where Høeg doubles down on the alien encounter by employing that well-trodden cliché of B-movie sci-fi horror: namely that the meteorite turns out to be riddled with strange parasites that threaten human life as we know it were they to be transported back to civilization where they could spread freely.

Set in the Antarctic rather than the Arctic, John W. Campbell's earlier short novella 'Who Goes There?' (1934) displays many of the elements employed in Høeg's novel. Adapted into the movie *The Thing from Another World* (1951), but gaining lasting fame via John Carpenter's adaptation of the story in his body horror classic *The Thing* (1982), Campbell's original story follows a group of scientists on a remote Antarctic base who discover 'a ship,' a 'thing like a submarine' that 'came down out of space.'[24] Rather than build suspense over several hundreds of pages as Høeg does in *Smilla*, Campbell almost immediately reveals not just the spaceship in which the alien visitor has arrived, but the pilots of the craft too, 'shadow-things'[25] encased in the ice that are clearly not human. As in *Smilla*, one of the scientists pleads caution, 'afraid that we may release a plague – some germ disease unknown to Earth – if we thaw those microscopic things that have been frozen there for twenty million years'.[26] This warning is however brushed aside as the scientists decide to haul a specimen back to base and thaw it. Unsurprisingly, things do not go as planned. The 'unearthly monster', a 'hellish creature' that is of 'an unearthly disposition' not only wakes up but immediately starts devouring the hapless scientist with 'an unearthly, unkillable vitality' and a 'furious monstrosity'.[27] In Campbell's story, but particularly in Carpenter's adaptation of Campbell's prose into visual format, the central monstrosity of 'the thing' is not its alien visage but its ability to mimic and hide among the human, as human. In the beginning of the novella as in the film, the initial contact with the alien life form elicits disgust as well as terror, but the true horror of the novella and (in particular) the film rely on the juxtaposition of the inherent monstrosity of 'the thing' in that it 'can imitate anything – that is, become anything',[28] yet in key moments reveal itself in grotesque distortions of the human and animal forms it otherwise so precisely replicate. The central characteristic of 'the thing' that is the monstrous is thus that there is no actual 'thing' in itself at its core, only veil upon veil upon veil, an endless series of approximations of human and animal forms that prove hollow.

In Høeg as in Campbell, the malignant visitor from outer space, whether it is that of a meteorite or a spaceship, a mysterious virus, parasite, alien or monster, is not exactly a novel trope. The space invasion narrative is one of, if not indeed, *the* most popular of science fiction tropes, ranging from literary classics like H.G. Wells's *The War of the Worlds* (1898) on to Liu Cixin's *The Three-Body Problem* (2014, orig 2008), but it is particularly prevalent in cinema. Films like *The Body Snatchers* (1956), *They Live* (1988), *Independence Day* (1996), *Mars Attacks!* (1996), *Signs* (2002), *Edge of Tomorrow* (2014), *A Quiet Place* (2018) and literally hundreds if not thousands of other films testify to this. As films like *Predator* (1987) or *Cloverfield* (2008) illustrate, any kind of environment can arguably serve as backdrop for the invasive force, whether it is that of a sweltering jungle (*Predator*) or a metropolis like New York (*Cloverfield*). World cities like Paris, London, Tokyo, Los Angeles, Moscow, Hong Kong, Rio de Janeiro or Sydney in particular serve as collateral in space invasion narratives displaying the violence and existential threat posed to humanity as such precisely by presenting the viewer with an apocalyptic smorgasbord of the world's most densely populated areas snuffed out in a matter of seconds. Still, the initial contact, as in the small, fictional desert town of Perfection, Nevada in *Tremors* (1990) or the real desert of Agua Dulce, California in *Nope* (2022), is often made in a space that, like the Arctic or the Antarctic, are perceived as underpopulated, spare and often inhospitable to human presence.

In this, the frozen deserts of the polar regions take up a familiar role to that of their hotter counterparts. As William L. Fox has observed in a range of insightful books on American deserts, 'big space means big time, the two being somewhat synonymous in our understanding of the universe.'[29] In the desert, geological, biological, cosmic and spiritual space and time intermix more freely than in other landscapes:

> This is a constant refrain you bump up against in the most arid, most absolute desert spots, from the bottom of the Sedan Crater, to here, to Meteor Crater in Arizona, to the Dry Valleys of the Antarctic: their use as analog environments for space exploration.[30]

Nevertheless, in their relative proximity to human presence and habitation, as in the sheer inhospitable nature to (human) life as such, the warmer deserts of the world and the Arctic cannot in the end hold a candle to the Antarctic. After all, as one of the characters of 'Who Goes There?' phrases it as he muses over the isolation of the remote scientific base, 'this is Antarctica, where there is not one single, solitary, living thing', nothing indeed 'except us' and the 'unhuman'[31] that walks among them. As an arena enacting a showdown between human and unhuman, civilized and barbarian, terrestrial and that which is not (extraterrestrial, subterranean, abyssal), of all the environments available on the surface of planet Earth, the Antarctic is routinely portrayed as the most alien, transgressive and otherworldly of spaces. Nowhere else on Earth is the passage to other worlds, other states of being (and to the otherness of other beings) as prevalent as in the deserted waste of the Antarctic.

Void, Abyssal, Chthonic: Monstrosity at the Mountains of Madness

In the polar regions, monsters, aliens and monstrosity are easier to commune with, but also more difficult to keep at bay, as is evidenced in many of the examples given here. Doctor Frankenstein's immoral act of creation, the creature itself and the violence following in its wake; the blinkered ambition and will of the officers of *The Terror*, the mysterious polar beast that terrorizes and haunts them, the acts of cannibalism; the dinosaurs of a hollow Earth in Verne's subterranean journey and the extraterrestrial alien invasiveness on both a macro-level (the meteorite) and micro (the parasite) of Høeg's tale of extraterrestrial visitation; and of course Campbell's and Carpenter's terror of the plasticity of a being that can mimic the human in its exterior manifestation while hiding its 'radically different'[32] nature in plain sight. As with the other deserts of the world, located far from the icy polar regions of the Arctic and the Antarctic, the apparent or actual lack of human, animal and vegetative life seems to open for the possibility of forms of life that is not only strange and otherworldly but also terrible.

It seems that the potential for life that is *inhuman, monstrous* and *alien* is imaginatively at its most vibrant, the more inhospitable and hostile an environment is to *human* life. And if the Arctic is 'like the desert in this way – open, unobstructed country', while also 'a wasteland of frozen mountains, of violent winds and incipient evil',[33] the Antarctic is doubly so. Further away from human civilization than any of the hot deserts of the world, all of which are at least skirted by human settlements and often interspersed with nomadic cultures, the Antarctic is far more desolate than the Sahara, the Gobi, the Mojave or the Taklamakan. As it is the case with the Arctic, the Antarctic is 'like' these places, yet also stands out in a range of critical ways. Defined as a negative, or perhaps more so a double (in that it stands in opposition to the Arctic but also echoes it), the Antarctic can etymologically and figuratively be viewed as a sort of enhanced mirror image of an environment that is already defined by emptiness, extreme inhospitality and a liminal porosity that engenders intimacy with the monstrous and the unfamiliar. Unlike the Arctic, placed by Eurocentric mapping at the top and with North as 'up', the Antarctic has traditionally considered to be at the bottom of the world. And as any child who have owned a globe will know: not only is the Antarctic routinely placed at the literal bottom of the globe, it is usually also partly obscured by the stand and fittings. In mapmaking as in globular projections, a Eurocentric perspective of the world has ensured that any close scrutiny of the place is not only much harder than its northern equivalent but also of lesser importance. It is not by accident that in the race for the poles, the South Pole came in second. Still, it is equally important that even though it was considered less significant than its northern equivalent, the eventual outcome and (for Scott and his fatal expedition) tale of the endeavour outshines the earlier and successful mission to reach the North Pole. In the monstrosity of its environment, in the situation that Scott and his men would find themselves in, and in the grotesque and painful manner in which they met their ends. Scott's tale of failure has become the most famous Polar tale of all.

In the kaiju[34] monster movie *Godzilla vs. Kong* (2021), the monstrousness of the land at the bottom of the world is merged with its utmost liminality and, in particular, with its antipodal position of being topsy-turvy. This is literally the case as the upside down of the Antarctic is played to full effect as the titular monster of the film, Kong, enters a hollow Earth in which even gravity is reversed. When the giant ape Kong is transported from tropical Skull Island in the Indian Ocean, it is taken to frozen Antarctica, a seemingly empty wasteland of ice and snow, but in fact turns out be hiding the gateway to a world that is literally in opposition to this one. As Kong enters a cavernous underground realm, he discovers not only that Earth is hollow, but that gravity can work in reverse directions. Furthermore, the hollow Earth discovered beneath (or above?) the Antarctic rapidly emerges to be the ancestral home not only to an entire species of giant apes like himself but to a multitude of giant beasts ranging from enormous crabs, arachnids and huge lizards, including the lizard above them all, Godzilla himself. In this, the movie follows in the footsteps of a long tradition of '"cryonic" fiction and time -travel narratives wherein the Antarctic landscape disrupts the normal linear functioning of time' and in which 'the Antarctic [itself] is represented as a landscape "out of time"'.[35]

The extreme remoteness, impenetrability, occulted vision, enormous monstrosity and a temporal perspective that has no regard for human history is exploited to its fullest potential by H.P. Lovecraft in his novella *At the Mountains of Madness* (1936). The tale of a doomed expedition to the Antarctic is described by the first-person narrator as an 'ultimate waste of forbidden secrets and inhuman, aeon-cursed desolation' and the source of 'unnamable and perhaps immeasurable evils'.[36] As readers familiar with Lovecraft will know, the prose as well as subject material of Lovecraft's writings are never not impenetrable and 'unnamable', but the tendency towards the unspeakable cosmic monstrosity of the world reaches a frenzied peak in his descriptions of the Antarctic. It is described as 'an accursed realm where life and death, space and time, have made black and blasphemous alliances in the unknown epochs since matter first writhed and swam on the planet's scarce-cooled crust', beings who had developed advanced civilizations when 'dinosaurs were new and almost brainless objects'.[37]

The Antarctic, as seen from Lovecraft's perspective, is thus not so much 'a landscape out-of-time'[38] as it is a dimension following another timeline altogether, one that is not only older than human history and humankind as such but entirely indifferent to it. Accordingly, in Lovecraft's vision, it ends up as scarcely of the Earth at all. Or rather this is an alternative Earth, a world populated by 'elder things' that are of cosmic origin and cosmic time, and thereby the one 'part of the world [where] there had been a remarkable and unique degree of continuity between the life of over three hundred million years ago and that of only thirty million years ago.'[39] As a place where forms and temporalities stretch over hundreds of millions of years rather than decades or centuries, individual human perspective, or even that of intergenerational purviews, wither and collapse in the face of a 'cosmic horror' that renders human life and purpose irrelevant. Void of human history but overwhelmingly present as a prism of geological and cosmic timescales, the Antarctic thus acts as a bulwark 'against anthropocentrism – the world as instrumentally made for the human, the world as a world for human use and benefit.'[40] In this, the Antarctic becomes a repository of inhuman and cosmic temporalities and ontologies in which the human has little if any traction, a place that is not so much against the human as absolutely indifferent to it.

For many critics, Lovecraft's desire to this cosmic horror into words via 'systematically debilitating content'[41] is for better or worse one of the central characteristics of a style that threatens, or promises, to take centre stage. 'No other writer is so perplexed by the gap between objects and the power of language to describe them,'[42] Graham Harman concludes in his *Weird Realism: Lovecraft and Philosophy* (2011), one of the main theoretical works in a recent surge of books that helped elevate Lovecraft from pulp hack to an intellectual cult phenomenon. Still, for all of Harman's attempt to exalt Lovecraft to 'one of the greatest [writers] of the twentieth century' (Harman 10) and a concurrent pop cultural resurgence of Lovecraft that has resulted in feature films like *The Colour Out of Space* (2019) and the critically acclaimed television show *Lovecraft Country* (2020), adapted from Matt Ruff's equally well-received novel of the same name, Lovecraft remains a difficult sell to many people and for a variety of reasons. Lovecraft's 'tendency to suggest that his fictional horrors are indescribable'[43] only to then proceed to describe them in great detail has been viewed by many critics as evidence of Lovecraft's ineptitude as a writer, as 'excessive, embarrassing, excrescences on the corpus of literature.'[44] Combined with a host of problematic opinions that render him as a person 'abominable in ethical and political terms,'[45] and a corpus of writing testifying to this, any contemporary reading of Lovecraft must reckon with the fact that his writing was 'spun from utterly toxic aspects of modernity.'[46] The racism, chauvinism and speciesism exhibited by Lovecraft were never easy to swallow to begin with and as New Weird author China Miéville has remarked, it is not enough to simply "acknowledge" that he was a racist [… who] had regrettable opinions' only to then proceed to 'geekily enthuse' about his merits.[47] As Miéville insists, one has to read this into every single aspect of Lovecraft's storytelling and the ideology that drives it forward.

With this in mind, it is however telling that it is in the Antarctic, of all places, that Lovecraft finds what would be one of his most sustained (one could also say long-winded) digressions on the Cthulhu mythos but to some extent also his most accepting. While not exactly devoid of problematic positions on evolution, miscegenation, hybrid cultures and 'that monstrous chapter of pre-human life',[48] *At the Mountains of Madness* is nowhere near the rabid and racist frenzy of the worst of Lovecraft's tales, as in for instance 'The Horror of Red Hook' (1925). Set in the Brooklyn neighbourhood of that name, what horrifies the narrator of 'The Horror of Red Hook' is the bustling crowd of (non-white) races he deems unsavoury, prompting him to spew a vitriolic streak of offensive language and commentary.

The narrator of the latter story is a thin veil mirroring Lovecraft's personal disgust at having to live in Red Hook among people he deemed inferior. In comparison, the much longer tale set in the Antarctic, far away from the New York slum, is by Lovecraft's standards muted. As Brian Johnson has pointed out, 'the curious reversal of narrative sympathies that Lovecraft effects in *At the Mountains of Madness*'[49] is an anomaly in the Lovecraft universe, with what must be considered a rare concession to the plight of the other. Uncharacteristically for a Lovecraft protagonist, the narrator of *At the Mountains of Madness* realizes not only that monstrous as the encountered beings might seem to him, 'they were not evil things of their kind [but simply] another order of being.'[50] Johnson goes on to claim that this '"posthumanist" tendency to decenter human identity by reconnecting it to often troubling or defamiliarizing images of nonhuman nature makes his poetics of the materialist grotesque a suggestive precursor to the new ecological materialisms.'[51] This is perhaps to stretch a Lovecraft reclamation project a bit too far. Nevertheless, *At the Mountains of Madness* arguably makes for a less uncomfortable read than Lovecraft's typical texts in the so-called Cthulhu Mythos, a range of stories centred around Cthulhu, a 'gelatinous green immensity,' a 'green, sticky spawn of the stars' so hideous that any human will be struck mad just at the sight of such 'eldritch contradictions of all matter, force and cosmic order.'[52]

The horror and 'madness' at the core of Lovecraft's Antarctic novella is different, still alien and grotesque, but more sedate and at times bordering on marvelling. As with the monstrous 'thing' of Campbell's novella, a being that is described as having 'a sort of enormously slowed existence' given that a spark of life has lingered in it 'since Antarctica froze twenty millions years ago,'[53] it is as much time and prehuman history itself – a cosmic enormity in which human history and presence is hardly even a blip – as it is any bodily manifestation of the monstrous that terrifies but also amazes in Lovecraft's vision of the Antarctic. As the narrator explains with wonder as well as dread: 'Their original place of advent to the planet was the Antarctic Ocean, and it is likely that they came not long after the matter forming the moon was wrenched from the neighbouring South Pacific.'[54] The Antarctic, the oceans, the moon in sky, and the sky itself end up interchangeable, a vast cosmic arena in which the human plays not part at all.

Coda: The Certain Death of Penguins

In *Encounters at the End of the World* (2007), director Werner Herzog assures his audience that what he set out to do with this documentary was to 'not come up with another film about penguins' (00:02:47–00:02:50). As is tradition with the Herzog brand, this journey to 'a seemingly endless void' (00:02:47–00:02:50) is used as a platform from which the great existentialist questions in life can be posed and not, as Herzog's disavowal of 'penguin films' seems to imply, the 'pristine landscapes [and] fluffy penguins' (00:09:32–00:09:37). This, as he assures us, is what documentaries on the Antarctic tend to revolve around. Ironically, the one scene that the film is now primarily known for is however, of course, precisely that. In an aside, Herzog briefly veers off on questions of whether or not animals can go insane, precisely as a solitary and 'deranged' penguin is caught on camera 'heading into the interior of the vast continent [and] towards certain death' (01:14:50–01:15:00). Set to somber choral music and Herzog's equally severe trademark diction, the puzzling and sad fate of the apparently suicidal penguin has become emblematic not just of this particular film, but of Herzog as a film maker as such, mimicked by hundreds of memes and videos both cherishing and ridiculing Herzog as an existentialist director of a bygone world that the future can no longer hope to entertain in any meaning of the word.

As explored in this chapter, the Antarctic has been envisioned as a space of otherness, opposition. It is a space where gateways open up to subterranean and extraterrestrial monstrosities, the place on Earth where aliens can hibernate for tens and hundreds of millions of years, undisturbed until they are not. It is a liminal and transgressive space and an environment that is impossible to permanently call home. Unless of course one is a penguin. Because for penguins, including deranged ones, there is obviously nothing out of the ordinary about the Antarctic. While auteur filmmakers like Herzog may baulk at making yet another film about penguins, it is telling that this is in fact exactly what he ends up doing. In this, it is important to address that penguins are as emblematic of the Antarctic as polar bears are of the Arctic. Yet also that while they arguably share a sense of 'fluffiness' with polar bears, the fluff of penguins is however of a different order to that of their Arctic cousins. Mildly ridiculous and waddlingly inept when on land, penguins do not predate on humans. On the contrary, they are generally presented for comedic effect or of unexpected care and cuteness in an otherwise uncaring landscape, the most extreme of the latter evidenced in the nature documentary *March of the Penguins* (2005).

Exceptions to this do exist, including Monty Python's giant electric penguins in the sketch 'Scott of the Sahara' (S 2, E 10), as well as a race of equally giant albeit unelectrified blind albino penguins that make an appearance late in *At the Mountains of Madness*. Yet even such supposed monstrous mutations of 'the natural' are invariably comedic, whether by intention (Monty Python) or not (Lovecraft). Still, as Herzog's suicidal penguin reminds us, they can also be tragic. Because even though neither the actual nor any perceived and mediated threat to the penguin can as yet be said to be anywhere near that of the polar bear, given a continued rise in temperature, it inevitably will be. At that point, the polar bears will long have been extinct and the desert, the warm kind, will likely rule the Earth. As one of Herzog's interviewees prophetically concludes it (in the mid-noughties when this was still up for debate): 'If you can come to the Antarctic, please do. Plus … be aware of global warming. It's real' (00:34:25–00:35:32).

In this, the perceived fortitude and harshness of the polar regions in general, and the Antarctic in particular, is put into play in a way that is retrospectively and figuratively chilling (if not literally so). This is especially the case when viewed against the backdrop of the current reality of the Antarctic a decade and a half after the release of

Herzog's film. At the time of writing, in what is currently the winter of the Antarctic but also the hottest year on record, the Antarctic sea ice is at a historic low. As scientists confirm what they have for a long time feared, and warned against, it is becoming resoundingly clear not just that the Antarctic is under threat, but also that the story of the frozen desert at the bottom of the world is changing and rapidly so. Once a place to be shunned due to its hostile and unforgiving nature, this otherwise long-lived trope is shifting as quickly as the ice is melting. Increasingly imagined as a territory in need of urgent need of care and protection, the Antarctic is becoming emblematic not so much of the cosmic horror of the eons of the stars or of monstrous life forms of extraterrestrial origin, but of the urgent need to fix the human horror of emissions that is insidiously gnawing away at the basis of all life: human, animal, vegetal.

Notes

1 In 2007, Russia undertook the spectacular move of going 'deep below the North Pole in a submersible [and] planted their national flag on the seabed to stake a symbolic claim to the energy riches of the Arctic.... Russia wants to extend right up to the North Pole the territory it controls in the Arctic, believed to hold vast reserves of untapped oil and natural gas, which is expected to become more accessible as climate change melts the ice.' (Guy Faulconbridge, 'Russian Sub Plants Flag Under North Pole', accessed August 25, 2023, https://www.reuters.com/article/idINIndia-28784420070802). The most blatant example of this newly awakened imperialist interest in the Arctic is however without a doubt that of former US president Trump's attempt to purchase Greenland from Denmark in 2019. As Trump stated it in an interview at the time: "You take a look at a map. I'm a real estate developer. I look at a corner, I say, 'I've got to get that store for the building that I'm building,' etc. It's not that different. [...] I love maps. And I always said: 'Look at the size of this. It's massive. That should be part of the United States.'" (Peter Baker, 'Cosmetics Billionaire Convinced Trump That the US Should Buy Greenland," accessed August 25, 2023, https://www.nytimes.com/2022/09/14/us/politics/trump-greenland.html).

2 Jonathan Raban. *Arabia* (Picador, 1987), 160.

3 H. P. Lovecraft. *At the Mountains of Madness*. In *H. P. Lovecraft: Omnibus 1 – At the Mountains of Madness* (Grafton, 1985), 138.

4 Francis Spufford. "Scott Dies." In *The Ends of the Earth, Vol. II: The Antarctic*, ed. Francis Spufford (Granta Books, 2007), 13.

5 Sinéad Moriarty. *Antarctica in British Children's Literature* (Routledge, 2021), 6.

6 Moriarty, *Antarctica*, 16.

7 Elizabeth Kolbert. "Introduction." In *The Ends of the Earth, Vol. I: The Arctic*, ed. Elizabeth Kolbert (Granta Books, 2007), 4.

8 Sara Wheeler. *The Magnetic North: Travels in the Arctic* (Vintage, 2010), 4. Sara Wheeler. *Terra Incognita: Travels in Antarctica* (Vintage, 1997), 3. Wheeler, *Arctic*, 4.

9 Kolbert, "Introduction," 2.

10 Moriarty, *Antarctica*, 135.

11 Mary Louise Pratt. *Imperial Eyes: Travel Writing and Transculturation* (Routledge, 2000).

12 Peter Davidson. *The Idea of North* (Reaktion Books, 2007), 13.

13 Roland Huntford. "Introduction." In *Farthest North: The Incredible Three-Year Voyage to the Frozen Latitudes of the North* (Modern Library, 1999), xv.

14 Fridtjof Nansen. *Farthest North: The Incredible Three-Year Voyage to the Frozen Latitudes of the North* (Modern Library: 1999), 4.

15 Nansen, *North*, 4.

16 Nansen, *North*, 4.
17 Nansen, *North*, 3.
18 Nansen, *North*, 21, italics in original.
19 Nansen, *North*, 98.
20 Mary Shelley. *Frankenstein* (Oxford University Press, 1980), 16.
21 Shelley, *Frankenstein*, 210, 27, 26, 206 and 24.
22 Shelley, *Frankenstein*, 30, 222 and 213.
23 Shelley, *Frankenstein*, 21.
24 John W Campbell. *Who Goes There?* (Orion, 1988), 4.
25 Campbell, *Who*, 7.
26 Campbell, *Who*, 9–10.
27 Campbell, *Who*, 23, 25 and 26.
28 Campbell, *Who*, 30.
29 William L. Fox. *The Void, the Grid & the Sign: Traversing the Great Basin* (The University of Utah Press, 2000), 6.
30 William L. Fox. *Playa Works: The Myth of the Empty* (University of Nevada Press, 2002), 107.
31 Campbell, *Who*, 31 and 37.
32 Campbell, *Who*, 33.
33 Barry Lopez. *Arctic Dreams: Imagination and Desire in a Northern Landscape* (Harvill, 1999), 23 and 17.
34 Kaiju, Japanese for "strange beast", originally referred to Japanese (e.g. *Godzilla* 1954, *Rodan* 1956) and later Asian giant monster films (e.g. *Bulgasari* 1962, *The Host* 2006). The term is however now commonly used as a denominator for all giant monster movies, including retrospectively for earlier Hollywood films (e.g. *King Kong* 1933, *It Came from Beneath the Sea* 1955).
35 Moriarty, *Antarctica*, 12.
36 Lovecraft, *Mountains*, 58.
37 Lovecraft, *Mountains*, 51 and 81.
38 Moriarty, *Antarctica*, 32.
39 Lovecraft, *Mountains*, 30.
40 Eugene Thacker. *Tentacles Longer than Night: Horror of Philosophy Vol. 3* (Winchester: Zero Books, 2015), 124.
41 Graham Harman. *Weird Realism: Lovecraft and Philosophy* (Zero Books, 2012), 23.
42 Harman, *Weird*, 3.
43 Harman, *Weird*, 29.
44 David Punter. "Lovecraft: Suspicion, Pattern Recognition, Paranoia." In *The Age of Lovecraft*, eds. Carl H. Sederholm and Jeffrey Andrew Weinstock (University of Minnesota Press, 2016), 188.
45 Harman, *Weird*, 60.
46 Jeffrey Andrew Weinstock. "Afterword: Interview with China Miéville." In *The Age of Lovecraft*, eds. Carl H. Sederholm and Jeffrey Andrew Weinstock (University of Minnesota Press, 2016), 241.
47 Weinstock, "Afterword," 242.
48 Lovecraft, *Mountains*, 82.
49 Brian Johnson. "Prehistories of Posthumanism: Cosmic Indifferentism, Alien Genesis, and Ecology from H. P. Lovecraft to Ridley Scott." In *The Age of Lovecraft*, eds. Carl H. Sederholm and Jeffrey Andrew Weinstock (University of Minnesota Press, 2016), 104.
50 Lovecraft quoted in Johnson, "Prehistories," 105.
51 Johnson, "Prehistories," 105.
52 H. P. Lovecraft. "The Call of Cthulhu." In *H. P. Lovecraft and others: Tales of the Cthulhu Mythos*, ed. August Derleth (Grafton, 1988), 47.
53 Campbell, *Who*, 33 and 4.
54 Lovecraft, *Mountains*, 89.

Works Cited

Campbell, John W. *Who Goes There?* Orion, 1988.
Davidson, Peter. *The Idea of North*. Reaktion Books, 2007.
Fox, William L. *Playa Works: The Myth of the Empty*. University of Nevada Press, 2002.
Fox, William L. *The Void, the Grid & the Sign: Traversing the Great Basin*. The University of Utah Press, 2000.
Harman, Graham. *Weird Realism: Lovecraft and Philosophy*. Zero Books, 2012.
Herzog, Werner, director. *Encounters at the End of the World*. Image Entertainment, 2007, 99 mins.
Høeg, Peter. *Miss Smilla's Feeling for Snow*. Vintage, 2012.
Huntford, Roland. 'Introduction.' In *Farthest North: The Incredible Three-Year Voyage to the Frozen Latitudes of the North*. Modern Library, 1999, xiii–xviii.
Johnson, Brian. 'Prehistories of Posthumanism: Cosmic Indifferentism, Alien Genesis, and Ecology from H. P. Lovecraft to Ridley Scott.' In *The Age of Lovecraft*. Eds. Carl H. Sederholm and Jeffrey Andrew Weinstock. University of Minnesota Press, 2016, 97–116.
Kolbert, Elizabeth. 'Introduction.' In *The Ends of the Earth, Vol. I: The Arctic*. Ed. Elizabeth Kolbert. Granta Books, 2007, 1–8.
Lopez, Barry. *Arctic Dreams: Imagination and Desire in a Northern Landscape*. Harvill, 1999.
Lovecraft, H. P. 'At the Mountains of Madness.' In *H. P. Lovecraft: Omnibus 1 – At the Mountains of Madness*. Grafton, 1985, 9–140.
Lovecraft, H. P. 'The Call of Cthulhu.' In *H. P. Lovecraft and Others: Tales of the Cthulhu Mythos*. Ed. August Derleth. *Omnibus*. Grafton, 1988, 17–50.
Moriarty, Sinéad. *Antarctica in British Children's Literature*. Routledge, 2021.
Nansen, Fridtjof. *Farthest North: The Incredible Three-Year Voyage to the Frozen Latitudes of the North*. Modern Library, 1999.
Pratt, Mary Louise. *Imperial Eyes: Travel Writing and Transculturation*. Routledge, 2000.
Punter, David. 'Lovecraft: Suspicion, Pattern Recognition, Paranoia.' In *The Age of Lovecraft*. Eds. Carl H. Sederholm, and Jeffrey Andrew Weinstock. University of Minnesota Press, 2016, 183–198.
Raban, Jonathan. *Arabia*. Picador, 1987.
Shelley, Mary. *Frankenstein*. Oxford University Press, 1980.
Spufford, Francis. 'Scott Dies.' In *The Ends of the Earth, Vol. II: The Antarctic*. Ed. Francis Spufford. Granta Books, 2007, 130–136.
Sykes, Brad. *Terror in the Desert: Dark Cinema of the American Southwest*. McFarland, 2018.
Thacker, Eugene. *Tentacles Longer than Night: Horror of Philosophy, Vol. 3*. Zero Books, 2015.
Weinstock, Jeffrey Andrew. 'Afterword: Interview with China Miéville.' In *The Age of Lovecraft*. Eds. Carl H. Sederholm and Jeffrey Andrew Weinstock. University of Minnesota Press, 2016, 231–243.
Wheeler, Sara. *Terra Incognita: Travels in Antarctica*. Vintage, 1997.
Wheeler, Sara. *The Magnetic North: Travels in the Arctic*. Vintage, 2010.

3

DESERTS OF DOOM, SPACES OF POTENTIAL

Representations of the Desert in the Australian Imagination

Lesley Hawkes

Introduction

In Australia, there is a central opposition of two desert imaginaries: doom versus potential. This continued opposition grew out of a violent masculine settler colonial history that refused to acknowledge existing Indigenous ownership and knowledge and or other marginal perspectives. Rather than accepting the lived reality of the landscapes they found themselves in, colonial explorers and settlers held on to their imagined ideas and beliefs about deserts. People who did have the knowledge and experience to survive in these desert places were ignored. These colonial imagined desert regions have had a long-lasting effect on the Australian understanding of these spaces. This chapter will focus on the opposing representations of the desert and how and why this opposition occurs. It will also reveal how these different representations can alter the embedded national myths that have for so long upheld outdated colonial ideals.

When European explorers failed in their attempts to cross Australia, they and colonial history blamed the desert rather than lack of understanding and unpreparedness. This blaming can be seen in the failed expedition of three of Australia's best known colonial desert explorers: Charles Sturt (English), Robert O'Hara Burke (Irish) and William Wills (English). Charles Sturt held a firm belief that there must be an inland sea at the centre of Australia and in 1844 he led an expedition searching for this sea but instead found the dry Simpson Desert. He was devastated about his failure as he was convinced there had to be water in the centre of Australia. In 1860 Burke and Wills set out to cross Australia from South to North but they both died from exhaustion and starvation. One member of their party survived (John King) because he accepted help from the Yandruwandha, the local Indigenous Peoples, who kept King alive for two and a half months until Alfred William

DOI: 10.4324/9781003397212-5

Howitt 'rescued' him. These events reveal the lack of preparedness by the colonial explorers, and they also reveal the Indigenous Peoples had the knowledge to survive and thrive in these desert landscapes, but this knowledge was seldom sought or valued. In her book *Desert, Nature and Culture*, Roslynn Haynes evokes this ambivalence of deserts: "Deserts also threaten the inner self. [...] [But] for many Indigenous peoples, desert is home."[1] Haynes notes that to honestly confront these desert spaces required a shift in understanding and perspective and colonial explorers were not willing to change their ideologies to accommodate this shift. Instead, they chose to try to dominate and make these regions fit into their known boundaries. Aboriginal author, Alexis Wright, writes,

> these are the stories that Australia chose to tell itself and wanted to believe about a land that was stolen from Aboriginal people. This is how the imagination works in storytelling narratives and sometimes these imagined stories become embedded in the national mind.[2]

Colonial desert stories about doom, death, and dread have become embedded in the national settler psyche. However, there are other stories that have pushed past these imposed colonial limitations and reveal the beauty, life, and complexities of Australian desert regions.

European colonisers believed the arid Australian landscapes were harsh, unforgiving, and unliveable. This belief was pushed forward into the public psyche by the constant retelling of stories in history and fictional works about colonial explorers from England, and Europe dying lost and alone in desert spaces. Patrick White, one of Australia's most recognised and awarded authors, based his 1957 novel *Voss* on the ill-fated expedition of Prussian explorer Ludwig Leichardt. In 1848 Leichardt attempted to cross Australia from Queensland to Western Australia but his party disappeared, and it is believed they died somewhere in the Great Sandy Desert, an area of about 400,000 square kilometres. White's novel draws on the excitement but unpreparedness of the European explorer to make such a journey. *Voss's* protagonist Johann Ulrich Voss imagines triumphantly crossing the desert and becoming a hero in the process. He sees himself as superior to the people and the land around him, and it is his chosen duty to rule over both. White writes extensively on the search for transcendence throughout the novel, but ultimately the explorers must face their failure to cope with the reality that surrounds them:

> He fell, straight, deeply into himself. It was not possible really, that anyone could damage the Idea, however much they scratched it. Some vomited words. Some coughed up their dry souls in rebounding pea-pellets. To no earthly avail. Out of that sand, through which his own feet, with reverence for velvet, had begun to pay homage, rose the idea, its granite monolith untouched.[3]

Voss realises too late his spiritual quest should have been more aligned with the actual landscapes that surrounded him, instead of his imaginings. The desert in Voss's mind was not the desert he encountered. Isabel Rawlins and Myrtle Cooper, while researching Australian writing on dirt, find that, Voss's "orientation is spiritual and transcendent however, White's treatment of the desert shows conceptual and corporeal boundaries between human and environment."[4] The desert landscape and the human explorer never reconcile, and the explorer dies. The German explorer is disappointed at the end to realise that he is human and not a spiritual entity: "Then I must return to those men, he would realize, and the thought of it was terrible. Nothing could have been more awful than the fact that they were men."[5] At the end of the novel, a statue is erected to commemorate Voss's valiant and heroic attempt. The reader knows there is nothing heroic or brave about this expedition and the team of men were lost, crazed and mutineers. Voss lays dead and beheaded somewhere in the desert. However, Voss does get his monolithic symbolism:

> Johann Ulrich Voss was by now quite safe, it appeared. He was hung with garlands of rarest newspaper prose. They would write about him in history books. The wrinkles of his solid bronze trousers could afford to ignore the passage of time.[6]

These lines show how the colonial imagined desert spaces continue through the celebration of false historical retellings in newspapers and in the form of public statues.

These and other failed expeditions have engrained in the Australian psyche a connection between the desert, failure, and death while at the same time celebrating the European men who attempted to cross and control it. Of course, Australia is not the only country where deserts are popularly perceived in this manner. As Aidan Tynan writes, "deserts can be and have often been regarded as the geographical correlate of death, places in which life's organic limits are tested."[7] Australia's colonial past has made the fear of the unknown more prevalent with European colonisers trying to map what they knew onto unknown landscapes. In Australia, people have continued to cling to the outside of the map, with over 80 percent of people living in Australia's coastal zone and even though the desert regions (Great Victorian Desert, Great Sandy Desert, Tanami Desert, Simpson Desert, and Gibson Desert) make up a large percentage of Australia's landmass only three percent of the population call these deserts home. Of course, there are legitimate reasons for this: lack of infrastructure, lack of employment, extreme climate, and lack of housing and education. However, there is also little desire by most people to live in these regions and sustained negative representations of the desert in the Australian imagination have only aided in the desert being seen as undesirable and the 'dead centre' or the 'dead heart' of Australia.

Geologist Michael Welland has written,

> for Australia, the nation's 'dead heart' or, as a Melbourne newspaper termed it, 'the hideous blank' embedded itself in the national psyche as an embarrassment to be conquered, developed, and settled with the idea of 'terra nullius' always in mind.[8]

'Terra nullius,' land belonging to no-one, was a legal principle implemented into Australian Law in 1835 by New South Wales Governor Richard Burke. This term has been consistently used to validate and justify the colonisers' taking of land. This continued disregard for Indigenous ownership and living on the land is evident in the fact the British decided to test their nuclear weapons in Maralinga and Emu Fields in South Australia, and Montebello Islands in Western Australia between 1952 and 1963. The British chose these desert sites because they believed or chose to believe nobody lived in these regions, but Indigenous Peoples live and have always lived on these lands. The knowledge that the Indigenous Peoples could have brought to the nuclear discussion was never sought and the desert regions were treated, by the British government, as a wasteland and testing zone, not places of home. The opening of the film *Maralinga Tjarutja* (2020) states, "To understand country you have to live on country"[9] and the Indigenous Australians understand the 'beauty of desert.'[10] To the Indigenous People these sites are home, not wastelands. Liz Tynan writes that the "damage done to Indigenous People in the vicinity of all these three test sites is immeasurable and included displacement, injury and death."[11] This example demonstrates that little had changed in the attitudes from the first days of colonisation to the mid twentieth century when it came to deserts and seeing them as undesirable places of home. Alexis Wright, when writing about Indigenous concepts of home highlights the colonisers' disregard for Aboriginal understanding of the term,

> To Aboriginal people, the term outback is an alien concept that implies someone is a long way from home, and home is elsewhere. We see our country as our home whether it is in the desert, the forest or by the sea.[12]

'Outback' is a term used in Australia to signify you are a long way from the urban and built environment, and thereby a long way from where most people call home. In 1992, after a long legal battle to have Indigenous Land Rights recognised, terra nullius was finally overturned in the High Court of Australia.

The lingering effects of colonialisation can be seen in contemporary films such as *Mad Max* (1979) where deserts are represented as undesirable, inhospitable terrain, casting the environment as a post-apocalyptic setting and illustrating that in modern portrayals the desert is still imagined as early Europeans understood it—a wasteland. In some contemporary films, the desert becomes intrinsically linked with crime as in the 1988 *Evil Angels* movie (known overseas as *A Cry in the Dark*) on the death of Azaria Chamberlain. Azaria Chamberlain was a baby who went missing while her family was camping near Uluru. Her family maintained that a dingo had taken the child, but the mother was found guilty of the baby's murder. It was over 20 years before the court found that Lindy Chamberlain was innocent, and a dingo had taken the baby. Throughout the trial and in the film, it is continually questioned why a family would take a baby to this desert region for a holiday and there must have been a sinister motive in visiting such landscape. And the highly successful horror film *Wolf Creek* (2005) caused many travellers to fear breaking down in the desert in case they encountered a serial killer. Deserts are seen as places of lived nightmares or fields of nothingness. The deserts of Australia remain the great unknown causing colonising attitudes and imagined fears to continue. However, there are other representations of Australian deserts that are not just places of violence but of life as well.

More recently, there have been cinematic and textual representations that have fought back against these negative stereotypes to present the desert landscapes as places of promise, potential, and home. Works such as *Tracks* (1980), *Rabbit-Proof Fence* (2002), *The Tracker* (2003), *The Oldest Song in the World* (2012), and *The Shepard's Hut* (2018) are textual and filmic examples that show the desert as complex and diverse places offering up opportunities for renewal and acceptance that other landscapes cannot. Claiming back the potential of deserts is a way to write back against the colonising of the land and culture and to move past the obsession that Western storytellers have with the pastoral mode and the colour 'green.'[13] Scott Slovic, in his book *Getting Over the Colour Green* (2001), finds that it has taken American literature a long time to get over the colour green and the same is true in Australian literature. These desert spaces offer up places for new narratives to be heard. Desert stories are often told by people who have been marginalised by more Western stories and the desert spaces become places that enable these diverse stories to be shared. They also become spaces where human and landscape become reconciled with each other. These desert stories write back against the dominant representations of arid landscapes and reveal the complexities and beauty within these spaces.

Desert Doom Aesthetics

The *Mad Max* (beginning 1979) series of movies, directed by George Miller, are some of the most recognisable cinematic representations of Australia. These movies conjure up images of deserts and post-apocalyptic landscapes where people with opposing views of humanity fight to death. As this series of films progress the desert as setting comes to play a greater role. *Mad Max Fury Road* (2015) begins with Max standing in front on his car in the desert. Over this image comes a narration: "My name is Max. My world is oil and blood… It was hard to know who was more crazy me or everyone else."[14] The films are essentially about oil and its scarcity and what this scarcity leads men to do. Max is a broken man in a broken car in a broken landscape. *Mad Max Fury Road* (the fourth instalment) was meant to be shot in Broken Hill, but the desert regions of that area had received a lot of rain and the wildflowers had turned the desert into a picture-perfect setting. After rains the Australian desert plains become sites of regeneration, beauty, and growth. However, this is not what comes into the mind when thinking of a desert. In this instance, the actual landscape did not fit in with notions of the 'imagined real' desert and the planned film location was rejected. People find it hard to accept that deserts can be filled with colour and flowers. Edward Said writes of 'imaginative geography' and how certain landscapes are imagined into reality while the lived environment is neglected as 'unreal: "The book (or text) acquires a greater authority, and use, even than the actuality it describes."[15] If *Mad Max Fury Road* was shot in the 'real Australian' desert while it was filled with wildflowers, it would not have fit in with the imagined landscape of the Australian desert and audiences would question its legitimacy. It was decided to shoot the film in Namibia instead. What the relocation of the film setting reveals is the construction of an imagined Australian desert landscape and the effect this imagined landscape has on the understanding of desert spaces. Australian deserts are diverse and complex. However, many films try to universalise deserts and fit them all into one neat category. Tom Lynch writes that much writing "portrays the desert as a more-or-less undifferentiated blankness."[16] As the film makers of *Mad Max* discovered, the desert regions around Broken Hill are not always empty, dry, or colourless. However, the film makers decided that the desert in the film would 'look' more authentically Australian if it was shot in Namibia in South-West Africa devoid of any colour or life.

The 2005 film *Wolf Creek* cemented in the psyche of many travellers a deep fear of the dangers hiding in the desert. Mick Taylor, the psychopathic murderer in the story, symbolises the evil that awaits innocent and unprepared tourists if they venture into desert spaces. *Wolf Creek* has two opposing landscapes: the beach which is portrayed (in Broome and Sydney) as civilised, mapped, and celebrated and the opposing desert, which is alien, isolating, and to be feared. In the film, three young British tourists set out for a scenic trip to take in the enormity of the Australian outback, especially Wolfe Creek National Park and the huge Meteorite Crater within the Great Sandy Desert, Western Australia. Their car breaks down in the desert and it is Mick Taylor who stops for them and asks, "What the bloody hell are you doing out here?" From these opening words, it is implied that the young tourists had no right venturing into the desert. The young urban Sydneysider male is the only survivor from the trip. The young female foreign tourists are killed in the strange environment inhabited by unhinged males. Haynes writes,

> Deserts of all kinds have exerted a hold on the Western imagination, as expressed in fiction and film. In these media, the desert functions as code for extreme physical conditions, for isolation and introspection, or for the place where the expectations of civilized society no longer hold.[17]

This is very much the case in this film. Mick, who lives alone in the desert, has no respect for societal rules. The young city tourists are condemned to a harrowing death at the hands of this lawless character. Mick Taylor is the ultimate violent masculine figure. John Stratton observes that in "Australian mythology the drier the outback, the more lawless and threatening it is."[18] The film implies that Mick

Taylor has managed to live in this lawless desert region because he too is lawless and violent. Anthropologist Deborah Bird Rose reminds us,

> settler Australians understand the continent by reference to the distance between the centres (usually cities), and the remoteness of areas from cities. Much of the continent in settler terms, is remote. But for Aboriginal people who live in the remote areas (and undoubtedly for many settlers who also live there), where they live is the centre.[19]

In the film, there are no references to Indigenous life in the desert, so the desert is always a long way from any known 'home.' Tom Drahos says that the lack of Indigenous characters in the film and the lack of any references to their knowledge or their ownership of the land helps in promoting the desert as uninhabited and feared.[20] The film implies that nobody lives successfully in these desert regions except the madman Mick Taylor. Once again, the desert region is set up as a place where nobody of right mind would choose to call home and if they do then there must be a sinister reason behind it.

Real Desert Spaces

As previously stated, not all artistic representations of the desert are negative and there are stories that push back against the constant negativity associated with these regions. There are stories that stress the possibilities that deserts provide, and it is these diverse representations that this chapter will now focus on. *Tracks*, a best-selling memoir, by Robyn Davidson and later a successful film based on the memoir are prime examples of positive representations of the desert. In 1977, Robyn Davidson, a young Queenslander from the Darling Downs regions, walked across the Australian desert with three camels and a dog. It took her nine months to complete, and she kept a journal of her experience. *National Geographic* asked her to write an article for their magazine and a photographer met with her three times throughout the adventure to take accompanying photos. This article was such a success with the reading public that Davidson turned her journal into a memoir. She was a young, inexperienced female adventurer and did not fit the mould of the rugged Australian masculine explorer that Australians were used to reading about. The film, based on the memoir, opens with the words, 'Alice Springs, last Stop,' implying that anything past Alice Springs is the end of the line. Davidson alights from the train and begins her training of getting ready for the trek. In the film when Davidson tells people of her plans she is constantly told, "Don't die out there." However, Davidson feels a pull to go on with her adventure. In an interview with ABC Radio National (8th December) Davidson says, "She had a fascination with the desert and wonders now if those early sensual signals of dry air and smell of dry grass of her childhood ran deep."[21] She goes on to say, "Perhaps, all Australians have some sense of the desert back there buried in their psyches."[22] Throughout her trip Davidson says she became conscious of

> an awareness and an understanding of the earth as I learnt how to depend upon it. The openness and emptiness which had at first threatened me were now a comfort which allowed my sense of freedom and joyful aimlessness to grow.[23]

Davidson spent two years learning about camels and landscape in preparation for her travels across the desert.[24] She was also aided in her travel by an Indigenous elder for one month of her trip. She says that he shared with her the need to let go 'of the Western previous preoccupation with time and structure' in the desert region.[25] This awareness of her surroundings is in stark contrast to the character of Voss. He never learns to depend on the land and constantly fought against it, trying to prove he was master. Davidson finds that the sense of huge uninhabited space works deep in the Australian collective consciousness. In her memoir she finds that for

> many people in Australia, it is frightening and most of the people huddle around the eastern seaboard where life is easy and space a graspable concept, but it produces a sense of potential and possibility nevertheless that may not exist now in any European country.[26]

Davidson is drawing attention to the enormity of the spaces of the Australian deserts and instead of seeing these spaces as frightening she says they should be seen as spaces of possibility. She uses the word here as great in the moral sense not just large or vast in the geographical sense.

While *Mad Max*, *Wolf Creek*, and *Evil Angels* all have the car as the safe mode of transport across the desert, Davidson walks with her camels. When the characters in each of these films leave their cars, they find themselves in serious danger. And in the classic Australian film *Walkabout* (1971), the father destroys the safety of the car by setting it alight before killing himself and leaving his two children to walk through the desert. The films that display more complexities on deserts have walking very much at the centre of their narratives. The landscape and the character become more connected through the act of walking. Rawlins and Hooper find Robyn Davidson's "immersion in the dirt of the desert serves as a kind of communion with it that brings her physical and spiritual release."[27] There is a scene in the book where Davidson feels a strong connection with the land to the point where she realises, she must change her entire outlook of life:

> the self did not seem to be an entity living somewhere inside the skull, but a reaction between mind and stimulus. And when the stimulus was non-social, the self had a hard time defining its essence and realizing its dimensions. The self in a desert becomes more and more like the desert.[28]

She goes on to write,

> When you walk on, sleep on, stand on, defecate on, wallow in, get covered in, and eat the dirt around you, and when there is no one to remind you what society's rules are, and nothing to keep you linked to that society, you had better be prepared for some startling changes.[29]

This is the opposite of what happens in Patrick White's *Voss*. Johann Ulrich Voss never reconciles with the land around him and sees it only as a place for him to physically and spiritually fight against:

> This devilish country, flat at first, soon broke up into winding gullies, not particularly deep, but steep enough to wrench the backs of the animals that had to cross them, and to wear the bodies and nerves of the men by the frantic motion that it involved.[30]

Davidson embraces her surroundings, and it is the act of continued walking that enables her to feel this closeness. Voss resents his surroundings and the dangers involved in riding through it. Mick Taylor, in *Wolf Creek*, has left society's rules behind him, but he still believes he is the dominant individual in the environment rather than part of it. He uses his knowledge of cars to relay this control to his victims. Rebecca Solnit finds, "walking, ideally, is a state in which the mind, the body, and the world are aligned, as though they were three characters finally in conversation together."[31] This connection through walking and desert spaces can be seen in other positive representations of the Australian desert.

Rabbit Proof Fence: Desert as Home

The 2002 film *Rabbit Proof Fence*, an adaptation of a novel written by Aboriginal author Doris Pilkington Garimara, and directed by Phillip Noyce, differs in many ways from conventional representations of the desert. In this film the desert is home. It is the place that is longed for and dreamt about by the young protagonists, three young Aboriginal girls who are forcibly stolen from their families and taken to a mission school. The girls escape and make the long way back home. They are walking towards 'real desert country' not trying to escape the desert, but rather return to it. Their story tells of their long walk: "'Long way' sums up rather understatedly what was, without doubt, one of the longest walks in the history of the Australian outback."[32] The girls' identity and sense of belonging is bound up in the desert region. Their stories, their families, and their home are all linked to the desert regions. The desert is not viewed as the 'dead centre' or 'the dead heart' rather the desert regions, and the rabbit-proof fence that ran through it, was a 'symbol of love, home and security.'[33] As stated, the film is based on the 1996 book

Follow the Rabbit Proof Fence, a memoir, by Doris Pilkington Garimara and in the book she describes the bush rather than the desert as "A strange, scary wilderness to these three girls who came from the desert regions of Western Australia."[34] One of the most striking aspects of this story is the strength the young girls gain from being in their desert home region: "Both girls took in the familiar landscape of the red earth, the dry spinifex grass and the grey-green mulga trees. There was nothing to compare with the beauty of these plains that stretched out in all directions."[35] This example demonstrates how the young Indigenous girls are not bound by pastoral notions of green landscapes rather their interpretation of beauty comes from the lived experiences of these desert regions and the love of home. In *Rabbit Proof Fence* the desert provides safety for the girls. In *Evil Angels* the desert became the place where innocence was destroyed but in *Rabbit Proof Fence* it is the place where innocence is restored. Films *Mad Max* and *Wolf Creek* follow the colonial tradition of the desert as a place for violent men while stories such as *Tracks*, *Rabbit Proof Fence* and *The Oldest Song in the World* explore how it is women who understand and survive the desert regions. The women do not try to change the desert or overcome it but rather they learn to live within it. It is their home and where they feel safe.

The Oldest Song in the World: Letting go of Colonial Myths

The Oldest Song in the World by Sue Woolfe (2012) is a contemporary novel that tells the story of Kate, a mature age university student who travels to the Australian desert to undertake field work for her studies in linguistics. Her task is to record an Aboriginal ancient song so it can be analysed by academics: "This mission will turn me into the perfect student."[36] Once she arrives, she realises her imagined aesthetic of what desert life would be like did not match her lived reality. Alexis Wright explains how this imagined aesthetic comes into being,

> the building of an Australian nation required myths about resilience to build a stoic psyche for surviving a hard, hot, mostly drought-stricken land, popularised by the idea that this country needed to be ruthlessly conquered, tamed and exploited for its natural resources. This was the colonial romance.[37]

Kate thought she could go into the desert, get the story, and leave. She wanted to 'use' the land and its people to further her city career. However, the lived experience is not what she imagined:

> But now I had to cope with being alone in the desert at night. This desert wasn't ordinary, accommodating earth to be trampled carelessly underfoot. Humans would only ever be a tiny part of this desert's life and like a river in a running tide, it must be approached with a knowledge of its ways.[38]

Kate begins to realise that she had a very shallow understanding of the complexity of the desert:

> As I walked, it was as if I'd never seen beauty, this or any other. The desert and its people had reached into the new chambers of my heart and pushed them wide open, I could almost hear the swinging open of the doors, as if I had to learn over and over how beautiful the world's possibilities were, as if I hadn't been concentrating hard enough before.[39]

She also begins to understand that she must reconcile her beliefs with other possibilities, "but then I stopped talking to you, because I needed to drop into silence. The desert was singing to me."[40] Again, we see the protagonist reconciling with the land around her and letting go of imagined desert aesthetics to engage with the reality of the desert spaces around her. Woolfe has a similar experience to Davidson in that she accepts the landscape for what it is and rather than trying to change it to suit her, she changes because of her experiences in it.

The Shepard's Hut: Desert as Safe Space

The Shepard's Hut is a 2018 novel by Western Australia's much awarded and much read author Tim Winton. Winton is known as a beach writer and for his blending of local landscapes with marginalised characterisation. In this novel, Winton leaves the beach behind but keeps the sand as the story takes place on the salt flats of Western Australia. The protagonist in this novel is Jaxie Clackson, an abused teenage character who escapes to the desert on foot after fleeing the scene of a terrible accident in which his abusive father died. Knowing that he would be a suspect, Jaxie travels as far as he can from his small rural town. On his travels in the desert, he meets an old man, an ex-Irish priest, Fintan MacGillis, who is hiding from a past event that Jaxie never fully understands. The event is based on financial fraud on a huge scale. The story ends with a violent encounter with drug dealers where the old man is violently killed. After enacting revenge on the killers Jaxie realises his place in life: 'An instrument of God.' There are many places throughout the story where the landscape is harsh and endless. Yet, the landscape turns out to be more trustworthy and dependable than many of the humans who try to make a quick dollar out of it. The drug dealers only see the desert region as a place to be unseen and where they can grow drugs out of view. They have no connection or interest in the desert. There is a line from A.D. Hope's well-known poem *Australia* that reads, "Hoping, if still from Deserts the Prophets come." This line is also the title of a much read and cited book by academic Geoffrey Serle on the Australian creative spirit: *From Deserts the Prophets Come* (Serle, 1973). Winton's story brings this much-repeated sentence into a contemporary setting. A young male searching for a quiet life but continually being drawn into a violent world by older male characters.

This is a coming-of-age story, but with a difference. Jaxie is trying to escape violence, but violence is everywhere around him. It is the desert that offers him peace and space for reflection.

The desert becomes the dependable space where Jaxie can find solace and acceptance. He is still confronted with a violent reality, but the violence is man-made and brought into the landscape by people who are exploiting the desert with the sole purpose of economic gain. Jaxie's abusive father is always trying to make a fast dollar through illegal activities; the drug dealers and the ex-priest (in hiding from his past fraudulent activities) are also trying to make big money quickly and they all die violent deaths. All these characters have been driven by the pursuit of a fast dollar but the desert reveals this will never last and sooner or later this approach to life will die but the desert remains. Alexis Wright finds this search for economic gain through the exploitation of desert natural resources feeds into the popular myth that colonisers tell themselves about the need to show dominance through exploitation.[41] Fintan has lived in the desert for many years and teaches Jaxie to slow down and connect with his surroundings: "Have you not seen it crawl upon itself and fill and empty in the afternoons? Now and then, you know, I see the eastern shore in the sky, all those stunted trees standing on their heads above the desert."[42].... "And it's a rare and beautiful place, don't you think? With a memory. Sometimes I think perhaps it's all memory."[43] This description is similar to that in *The Oldest Song in the World* where the character sees the landscape as being alive and having a memory. Fintan says,

> everything that ever happened here is still present now. In the crust, underneath, in the vapours. These days I look and there it says to me: Here I am, son, still here. I was here before the likes of you and yours were born. Before you even drew breath, I am.[44]

Tynan finds, "the desert retains the trace of everything that might populate it and thus manifests infinite possibility."[45] The desert has witnessed the histories of many lives and experiences, and it holds these memories within it, sharing them with those who are willing to listen and learn from them. Young Jaxie does not try to make a profit from the desert, nor does he try to exploit it. Instead, he finds a new sense of self and belief system to live his life by. Unlike Voss, Jaxie does not look to the religious world for validation. Lyn McCredden and Nathanael O'Reilly in *Tim Winton: Critical Essays* (2014) find that Winton's writing is filled with 'the transcendent world sensed in the palpable and everyday.'[46] Jaxie finds redemption and hope in the desert regions of Western Australia.

The Tracker: The Desert Filled with Life

The Tracker (2002) is an award-winning film directed by Rolf de Heer and set in Arkaroola Sanctuary in the Flinders Ranges in South Australia. However, throughout the film, the region is never directly named. The film begins with an image of an Australian desert and the words 'somewhere in Australia.' It takes place in the historical period of 1922 and tells the story of an Aboriginal tracker, played by David Gulpilil, and three non-Indigenous lawmakers on the search for an Aboriginal man accused of murder. The deeper the non-Indigenous men go into the desert the more the Indigenous man takes control and gains power from his surroundings. The tracker is forced by authorities to trace the steps of the fugitive. Gulpilil (in later life his name Dalaithngu was used in all his productions) had acted in the iconic 1971 Australian desert movie *Walkabout* (directed by Nicolas Roeg). *The Tracker* is almost a writing back to this 1971 film, and while it still focuses on settler/Indigenous interactions, it shifts the power balance from settler perceptions of land to Indigenous knowing of land.

As earlier discussed, *Walkabout* tells the story of two children who must walk through the desert after their father torches their car and kills himself. The character Gulpilil plays in *Walkabout* is a mysterious figure who helps the lost British children by leading them safely through the desert but then he takes his own life. The Indigenous character had served his purpose by leading the children to civilisation, but once they reach populated areas his knowledge is no longer valued: "the Australian desert and bush were part of that savage land associated with savage people, and both were to be tamed thanks to colonization and assimilation."[47] In *Walkabout* it is implied that if the character, like the desert, cannot be tamed then he must disappear.

In *The Tracker* the Indigenous character is given far more agency and his knowledge becomes valued and recognised. The continued interaction between the desert landscape and the Aboriginal tracker changes the perspective of the youngest of the lawmakers. He begins to value the desert land and the knowledge of the tracker to the extent where he stops the leader of the group from murdering innocent Aboriginal men and women. The film weaves the landscape into the narrative in an almost quilt-like manner with the tracker crisscrossing the landscape, following the tracks and signs of the accused man but also following the tracks of native animals. The tracker highlights the multitude of life that is present in the surrounding landscape but unseen by the non-Indigenous lawmakers. It also highlights the colonisers' 'failure to grasp the reality of the country they purport to rule.'[48]

Conclusion

This chapter began with a discussion of two opposing viewpoints on the Australian desert: one of doom and the other of potential. One is shaped by colonial fear and masculine violence. The other emerges in opposition to this lingering and limiting perspective and is shaped by lived experiences. This is not to say that deserts are not difficult places to live in or to travel through but more the viewpoint of desert as potential gives the opportunity to hear the voices of those often marginalised by dominant stories. What these texts and films reveal is the diversity of the desert spaces. The characters in these stories do not pass unknowingly through these regions but rather they stop and dwell. A commonality in all these stories is the characters do not think they can control their environment rather they accept that they do not know all the knowledge that surrounds them, but they will try to learn. Tynan writes deserts, "challenge life's ability to make a place for itself. They thus tend to put our conceptions of place and belonging into question."[49] Deserts also make us rethink the conventional idea of 'home' and 'outback.' Sometimes, home can be in areas that are seen as 'outback' in mainstream representations.

Early colonial Australian representations of the desert, even imagined representations, tried to fit it neatly into preconceived notions and conventions while the discussed texts challenge these limiting approaches. The desert landscapes, just as the Indigenous Peoples had been, were colonised and their knowledge disregarded. This knowledge was not valued, and their desert stories were not heard. Colonial concepts of terra nullius fed into Australian national myths including quite recent cinematic stories about the deserts and home. However, there are stories that reveal deserts as valuable places and agential. Deserts are filled with their own stories, and their spaces reveal the diversity and complexity of life, but these spaces need to be seen on their own terms and continued limitations placed around these spaces need to be removed. The discussed desert stories begin to decolonise outdated desert aesthetics that were based on imagined colonial myths and bring to life the reality of the desert spaces. They demonstrate that even though these foundational national myths are embedded in Australia's psyche they can be challenged and shifted allowing new stories and imaginings to occur.

Notes

1 Laura Singeot. "The Swamp and Desert Tropes in Post-Apocalyptic Australian Indigenous Fiction: *The Swan Book* (2013) by Alexis Wright and *Terra Nullius* (2018) by Claire Coleman", *Commonwealth Essays and Studies*. Note 6, 13. [Online], 43.2 | 2021, Online since 23 July 2021, connection on 06 June 2022. URL: http://journals.openedition.org/ces/7448; https://doi.org/10.4000/ces.7448
2 Alexis Wright, "A Dream of Belonging: For Indigenous Australians the Fabled 'Outback' Home." *The Guardian*, 18th May, 2021, 1.
3 Patrick White, *Voss*. Random House, 1957, 39.
4 Isabel Rawlins and Myrtle Hooper, "'Journeys into Dirt' in Robyn Davidson's *Tracks* (1980) and Patrick White's *Voss* (1957)." *Literator: Journal of Literary Criticism, Comparative Linguistics and Literary Studies*, 43 (1), a1806, 1. https://doi.org/10.4102/lit.v43i1.1806 2022.
5 White, 336.
6 White, 445.
7 Aidan Tynan. *The Desert in Modern Literature and Philosophy: Wasteland Aesthetics*. Edinburgh University Press, 2020, 3. http://www.jstor.org/stable/10.3366/j.ctv1453jh0.
8 Michael Welland. *The Desert: Lands of Lost Borders*. London: Reaktion Books, Limited. Accessed June 8, 2022. ProQuest Ebook Central, 2014.
9 Larissa Behrendt (Dir). *Maralinga Tjarutja*. Australian Broadcasting Corporation, 2020, 0:03:10.
10 Behrendt, 0:03:33.

11 Liz Tynan, "Sixty Years on, Maralinga Reminds Us Not to Put Security Over Safety." *The Conversation*: Published September, 26, 2016. Accessed 1st August, 2022, 1. https://theconversation.com/sixty-years-on-maralinga-reminds-us-not-to-put-security-over-safety-62441
12 Wright, 1.
13 Wallace, Stegnar. *Where the Bluebird Sings to the Lemonade Springs*. Modern Library, Random House 1992, 2002, 54.
14 George Miller, director. *Mad Max: Fury Road.* Australia: Roadshow Entertainment, 2015.
15 Edward Said, *Orientalism*. First Vintage books edition. Vintage Books, 1979, 93.
16 Tom Lynch, "Literature in the Arid Zone" in Cranston, C.A. and Robert Zeller. *The Littoral Zone: Australian Contexts & Their Writers*. Brill: 2007, pp. 71–92, 73.
17 Rachel Haynes, *Desert: Nature and Culture*. Reaktion Books, 2013, 152.
18 John Stratton. "Dying to Come to Australia: Asylum Seekers, Tourists and Death" in *Imagined Australia: Receptions Around the Reciprocal Construction of Identity Between Australia and Europe*, International Academic Publishers, 2009, 80.
19 Deborah Rose, *Nourishing Terrains: Australian Aboriginal Views of Landscape and Wilderness*. Australian Heritage Commission, 1996, 35.
20 Tom Drahos, "The Imagined Desert." *Coolabah*, Vol. 11, 2013, 148–161, 148.
21 Robyn Davidson, *ABC Radio National Interview by Cathy Pryer*, 8th December, 2017, 0:17:00 https://www.abc.net.au/listen/programs/lifematters/robyn-davidson-reflects-on-40-years-since-tracks/9178492
22 Davidson ABC Interview, 0:17:03.
23 Davidson, ABC Interview, 0:14:03.
24 Davidson, ABC Interview, 07:42.
25 Davidson, ABC Interview 0:14:00.
26 Robyn Davidson, *Tracks*. Bloomsbury, 1980, 191.
27 Rawlins and Hooper, 7.
28 Davidson, *Tracks*, 193.
29 Davidson, *Tracks*, 193.
30 White, 339.
31 Rebecca Solnit. *Wanderlust: A History of Walking*. Penguin Books, 2001.
32 Doris Pilkington Nugi Garimara. *Follow the Rabbit-Proof Fence*. University of Queensland Press, 1995, 100.
33 Doris Pilkington Nugi Garimara, 86.
34 Doris Pilkington Nugi Garimara, Xii.
35 Doris Pilkington Nugi Garimara, 95.
36 Sue Woolfe, *The Oldest Song in the World*. Fourth Estate, 2012, 34.
37 Wright, 1.
38 Woolfe, 80.
39 Woolfe, 389.
40 Woolfe, 389.
41 Wright, 1.
42 Tim Winton, *The Shepard's Hut*. Farrar, Straus and Giroux, 2018, 150.
43 Winton, 152.
44 Winton, 152.
45 Tynan, *The Desert in Modern Literature and Philosophy*, 4.
46 Lynn McCredden and Nathanael O'Reilly. *Tim Winton: Critical Essays*. UWA Publishing, 2014, 4.
47 Singeot, 1.
48 Wilson in Fiona Probyn. "An Ethics of Following and the No Road Film: Trackers, Followers and Fanatics." *Australian Humanities Review*, 37, 2005, 1. Accessed 27th June, 2023. https://australianhumanitiesreview.org/2005/12/01/an-ethics-of-following-and-the-no-road-film-trackers-followers-and-fanatics/
49 Tynan, *The Desert in Modern Literature and Philosophy*, 2.

Bibliography

Behrendt, Larissa, director. *Maralinga Tjarutja.* Australian Broadcasting Corporation, 2020. 52 mins.

Davidson, Robyn. *Tracks.* Bloomsbury, 1980.

———. "ABC Radio National Interview by Cathy Pryer." 8th December, 2017 https://www.abc.net.au/listen/programs/lifematters/robyn-davidson-reflects-on-40-years-since-tracks/9178492

Drahos, Tom. "The Imagined Desert." *Coolabah*, Vol. 11, 2013, 148–161. https://revistes.ub.edu/index.php/coolabah/article/view/15593

Elia T. Ben-Ari, "Defender of the Voiceless: Wallace Stegner's Conservation Legacy." *BioScience*, Vol. 50, Issue 3, March 2000, 253–257. https://doi.org/10.1641/0006-3568(2000)050[0253:DOTVWS]2.3.CO;2

Gulpilil, Graham Tardif, Rolf De Heer, Gary Sweet, Julie Ryan, Archie Roach, Noel Wilton, Damon Gameau, and Grant Page. *The Tracker.* Madman Entertainment, 2003.

Haynes, Roslynn D. *Desert: Nature and Culture*. Reaktion Books, 2013.

———. *Seeking the Centre: The Australian Desert in Literature, Art and Film.* Cambridge UP, 1998.

Hope, A. D. "Australia." https://allpoetry.com/poem/8509645-Australia-by-A-D-Hope

Lynch, Tom. "Literature in the Arid Zone" in Cranston, C.A. and Robert Zeller. *The Littoral Zone: Australian Contexts & Their Writers*. Brill: 2007, pp. 71–92.

McCredden, Lynn and Nathanael O'Reilly. *Tim Winton: Critical Essays*. UWA Publishing, 2014.

McLean, Greg, director. *Wolf Creek.* Roadshow Entertainment, 2005. 1 hr., 35 min.

———. *Wolf Creek 2.* Roadshow Entertainment, 2014. 1 hr., 46 min.

Miller, George, director. *Mad Max*. Warner Brothers, 1979. 1 hr., 33 min.

———. *Mad Max 2: The Road Warrior.* Warner Brothers, 1981. 1 hr., 36 min.

———. *Mad Max Beyond Thunderdome*. Warner Brothers, 1985. 1hr., 47 min.

———. *Mad Max: Fury Road.* Warner Brothers, 2015. 2 hrs.

Noyce, Philip, director. *Rabbit-Proof Fence.* Roadshow Films, 2002. 1 hr., 33 min.

Pilkington, Doris, Nugi Garimara. *Follow the Rabbit-Proof Fence*. University of Queensland Press, 1995.

Probyn, Fiona. "An Ethics of Following and the No Road Film: Trackers, Followers and Fanatics." *Australian Humanities Review*, (37), 2005. Accessed 27th June, 2023. https://australianhumanitiesreview.org/2005/12/01/an-ethics-of-following-and-the-no-road-film-trackers-followers-and-fanatics/

Rawlins, Isabel and Myrtle Hooper. "'Journeys into Dirt' in Robyn Davidson's *Tracks* (1980) and Patrick White's *Voss* (1957)." *Literator: Journal of Literary Criticism, Comparative Linguistics and Literary Studies*, Vol. 43, Issue. 1, 2022, a1806. https://doi.org/10.4102/lit.v43i1.1806 2022.

Roeg, Nicolas, director. *Walkabout.* Umbrella Entertainment, 1971. 1 hr., 40 min.

Rose, Deborah. *Nourishing Terrains: Australian Aboriginal Views of Landscape and Wilderness*. Australian Heritage Commission, 1996.

Said, Edward. *Orientalism*. First Vintage books edition. Vintage Books, 1979.

Serle, Geoffrey. *From the Deserts the Prophets Come*. William Heinemann Australia, 1973.

Singeot, Laura. "The Swamp and Desert Tropes in Post-Apocalyptic Australian Indigenous Fiction: *The Swan Book* (2013) by Alexis Wright and *Terra Nullius* (2018) by Claire Coleman." *Commonwealth Essays and Studies* [Online], 43.2 | 2021, Online since 23 July 2021, connection on 06 June 2022. http://journals.openedition.org/ces/7448; https://doi.org/10.4000/ces.7448

Slovic, Scott. *Getting Over the Color Green: Contemporary Environmental Literature of the Southwest*. University of Arizona Press, 2001.

Solnit, Rebecca, *Wanderlust: A History of Walking*. Penguin Books, 2001.

Stegnar, Wallace. *Where the Bluebird Sings to the Lemonade Springs*. Modern Library, Random House 1992, 2002.

Stratton, John. "Dying to Come to Australia: Asylum Seekers, Tourists and Death" in *Imagined Australia: Receptions Around the Reciprocal Construction of Identity Between Australia and Europe*, International Academic Publishers, 2009.

Tynan, Aidan. Interview Christopher Watkin. Crosscurrents Ecology, September 9th, 2020. https://christopherwatkin.com/2020/09/09/interview-with-aidan-tynan-about-his-new-book-the-desert-in-modern-literature-and-philosophy/

———. *The Desert in Modern Literature and Philosophy: Wasteland Aesthetics*. Edinburgh: Edinburgh University Press, 2020. http://www.jstor.org/stable/10.3366/j.ctv1453jh0.

Tynan, Liz. "Sixty Years on, Maralinga Reminds Us Not to Put Security Over Safety." *The Conversation*: Published September, 26, 2016. Accessed 1st August, 2022. https://theconversation.com/sixty-years-on-maralinga-reminds-us-not-to-put-security-over-safety-62441

———. "This Black Smoke Rolling Through the Mulga': Almost 70 Years on, It's Time to Remember the Atomic Tests at Emu Field." *The Conversation* Published: May 4, 2022 6.08am Accessed 9th June, 2022. https://theconversation.com/this-black-smoke-rolling-through-the-mulga-almost-70-years-on-its-time-to-remember-the-atomic-tests-at-emu-field-181061

Welland, Michael. *The Desert: Lands of Lost Borders*. Reaktion Books, Limited. Accessed June 8, 2022. ProQuest Ebook Central. 2014.

White, Patrick. *Voss*. Random House, 1957.

Winton, Tim. *The Shepard's Hut*. Farrar, Straus and Giroux, 2018.

Woolfe, Sue. *The Oldest Song in the World*. Fourth Estate, 2012.

Wright, Alexis. "A Dream of Belonging: For Indigenous Australians the Fabled 'Outback' Is Home." *The Guardian*, 18th May, 2021.

4

"BLINDED BY THE LIGHT"

Exploring the Desert as Luminous Landscape

Claudia Rosenhan

The storied desert constitutes a space where material agency and material discourses intersect to fabricate vivid and eloquent stories inscribed on and with matter through geophysical processes.[1] For example, the stratigraphic sequence of time inscribed in rocks will eventually weather into quartz sand, which entangles light, space, and time into what Barad has termed "spacetimemattering."[2] The matter of the desert is folded into strata folded into light and colour folded into resistance to smooth time flows and pleated into narrative dimensions of meaning.

Space Time

In Don DeLillo's stochastic novel *Point Omega* (2010), the universal point of Zen consciousness, towards which everything in the universe descends, is gathered into the desert. There Richard Elster has retreated after a career strategising the US invasion of Iraq. He is visited by filmmaker Jim Finley, and Elster's daughter Jessie, who suddenly disappears "past the edge of conjecture" (101). Thus human agency ends and time stretches indefinitely.

> *a car on the horizon floating slowly into motion, rippled in dust and haze, as in a long shot in a film, a moment of slow expectation.*
>
> *Point Omega (101)*

No cinematic cutting away across different scenes to show movement and action – the filmmaker resigns to the desert "endlessly not there"[3] yet eternally present.

> *I walked back into the wash under the shallow line of sky and then stopped and put my hand to the cliff wall and felt the tiered rock, horizontal cracks or*

DOI: 10.4324/9781003397212-6

> *shifts that made me think of huge upheavals. I closed my eyes and listened. The silence was complete. I'd never felt a stillness such as this, never such enveloping nothing.*
>
> *Point Omega (117–118)*

The multitude of strata are knitted together at points that are constantly unpicked. It is a fabric interleaved but forever fraying.[4]

Time Matter

This constant fraying connotes the eloquence of the landscape. In Jim Crace's story of redemption in the desert, Miri, pregnant and abused by her evil husband Musa, tries to weave these fraying narrative strands of the landscape into her birthing mat.

> *weavers who set their looms in open ground have first to find the landscape's warp and weft, the shadow lines, the tracks, the spirit paths. The weaving and the landscape should concur or else the cloth would lose its shape. The wind, the water and the threads, the lines of scree, the strata of rock, the patterned strips of wool should run in unison and then the fabric would be true [...] No weave could match such stringy wind or cluttered light.*
>
> *Quarantine (103–104)*

Whilst the desert is resistant to Miri's weave, her mat nevertheless is "informed by sunlight" and "tested by the wind". The desert light, in continuous shifting shadow, tangles the distinct strata of the rock formations like an extra thread that wraps over the warp without disturbing the basic weave and thus pleats the thread of time and meaning into new patterns, each new movement being inherent in the previous. As Deleuze noted, "Sandstone or basalt landscapes are made up of colours and lines, but they are always in movement".[5]

Matter Time

The nameless, burned pilot, considered to be English because of his accent, and whose memories of the desert fizz like sparks of lightning as he lies convalescent in an Italian monastery near the end of World War II, also likens the desert to cloth.

> *the desert could not be claimed or owned – it was a piece of cloth carried by winds, never held down by stones, and given a hundred shifting names long before Canterbury existed, long before battles and treaties quilted Europe and the East.*
>
> *The English Patient (147–148)*

Yet the threadings of material reality into strata of sand are constantly dissolved and are blown by the winds as crystallised spacetime, indexed in the "English" patient's notebook as colourful sliding matter, as wind that "dresses in bright yellow walls", or as wind that blows red "dust as fire", a dust storm that is a "copper-tinted" sheet. These are velocities of constantly shifting meaning.

Time End

> *there were two periods of time when he could not move. At noon, when the shadow was under him, and at twilight, between sunset and the appearance of stars. Then everything on the disc of the desert was the same.*
>
> *The English Patient (265)*

Michael Ondaatje's narrative, like DeLillo's, is one of disappearance and dissolution, eventually culminating in the detonation of the bombs in Hiroshima and Nagasaki. Thus, the time frame of zenith and azimuth erases temporal distinction. At subsolar and subhorizontal reflections, the traveller loses perception. The desert is where time stands still/speeds up (to atomic oblivion). On this plane, light has no time reference at all.

Perihelion

Deleuzian philosophy has specific resonance for thinking about storied landscapes, as it considers landscape as sensory and immanent space that favours such velocities over the transcendent verticalities of place. It is "a plane of Nature," and Deleuze and Guattari's idea of "a plane like the desert"[6] invokes space-time without structure or boundary, nomadically inhabited by continuous movements of energy, such as the winds above. Their scattering mirrors the movement of photons through spacetime, also at once material and immaterial. It is, thus, no great leap to identify the plane of immanence as "photonic," and by the same token the desert as luminous and crystalline landscape.[7] The plane of immanence "is surrounded by illusions [...] by thought's mirages [... that] arise from the plane itself, like vapours from a pond."[8] The plane of immanence, therefore, is light, is desert, that knows only intensities and affect.

The photon as energy vibration, Berressem points out, cuts across the physical and philosophical divide, referencing Bergson's universe as constant disturbance, and looping back to Heisenberg's notion of light as "pure potentiality."[9] This stimulates the possibility of transmutability and complementarity and helps thinking with light as indeterminate and diffused intensity that brings things into focus, often through absence. The resonance of light in the desert is firstly as a solar haze, a luminosity that "itself becomes opaque."[10] This "blurred vision" is concentrated through sand and sky into intensive colour, fiery reds and copper-tinted sheets of atmosphere into perception.[11]

Shadow Light

> *out there, on the floor of the desert, there had been a noiseless but almost explosive transformation. The clouds had shifted and the sun was now shining on*

> *the nearest of those abrupt and jagged buttes, which rose so inexplicably, like islands out of the enormous plain. A moment before they had been black and dead. Now suddenly they had come to life; between a shadowed foreground and a background of cloudy darkness they shone as if with their own incandescence.*
>
> *Ape and Essence (11)*

In Aldous Huxley's post-nuclear dystopia, William Tallis, the author of the eponymous screenplay, had lived and died in Murcia, California, "on the south-western fringe of the Mojave Desert." It is apposite that Huxley located his renegade author in the same desert he chose as a refuge, fascinated by its quality of light.[12]

However, the narrative of light and colours in the environment as bright sunshine or numinous moonlight is, for Huxley, "no more than names and notions, mere verbalizations, for utilitarian and scientific purposes, after the event."[13]

Light Spacetime

Instead, such phenomena should be considered as light "without knowing," "unmitigated Reality" or, in Deleuze's words "a Goethean moment."[14] In that sense, the desert is an unmediated landscape of light, not "emblematic botany." Light is beyond metaphor. Rather than acting as a lens for *reflecting on* the environment, light *entangles with* the perception of environment. The becoming of forms "trembling in the air" is epiphanic, a sudden revelatory instant of pure awareness.

> *sunset was a human invention, our perceptual arrangement of light and space into elements of wonder. We looked and wondered. There was a trembling in the air as the unnamed colors and landforms took on definition, a clarity of outline and extent.*
>
> *Point Omega (22)*

DeLillo shows how the initial clarity of outline slowly diffuses and dilutes spacetime again, until it is assimilated at the point of supreme consciousness. Time slows down the faster we approach the speed of light, until time stands still.

The sublimity of desert landscapes is undone under a relentless sun, scorched into a hallucinatory phantom full of intensities, that is, events actualised in the environment. The light-filled expanse of the desert acquires an affective dimension of space, silence and brightness. These intensities are actualised on the plane of immanence in a wave that moves "in one continuous stream of vibration" infinitely across the horizon.[15]

The Crystal-Image

Hanjo Berressem's exploration of Deleuze's "Luminous Philosophy" establishes light as the register through which a space may be viewed, as a "line of white light refracted by crystals."[16] Deleuze's "crystal-image" is, according to Berressem, the merger of the actual and the virtual image at a luminous point-at-infinity[17] and can help to understand the trajectory in which an image of the desert is no longer a reflection of what we think a desert is, but in which the light illuminating the desert landscape is a convergence of thought and materiality. Sight becomes "in-sight" that scintillates at different wavelengths.

Engraving "in"-sight into image via mechanistic reproduction onto light-sensitive matter is the realm of photography. The desert's crystalline reflective atmosphere, largely free from airborne pollutants, delaminates 2D photographic film and interweaves it into the fourth space-time dimension. Deleuze's writing on the cinema establishes the crystal-image as a time image of the present that splits into two: "one of which is launched towards the future while the other falls into the past."[18] Perceptions of past and future are enacted in the crystalline light, saturating with colour shifts as the waves stretch to the remote horizons.

Perception Colour

The conditions of perception, in Deleuze's words, become "when white darkens into yellow and black lightens into blue."[19]

> *the shadows are gulfs of indigo, and in the richly golden lights the far-off details stand out distinct and small and perfect, like the images of things in a convex mirror. In the foreground, delicately chased and stippled by the almost horizontal light, even the baldest patches of parched earth reveal an unsuspected sumptuousness of texture.*
>
> *Ape and Essence (127)*

The crystal-image of the desert as cinematic landscape is upended in Tallis's screenplay of humanity's descent into apehood, set in the man-made desert of post-nuclear California in 2108. The folded layers of heat and light bend in relationship with temperature and pressure across large distances. The decreasing light speed and the shortening of wavelengths of "almost horizontal light" scatter in the thickening atmosphere and redden the sun and plump the scorched stone of the desert, entangling light and heat. "The idea in space is vision" which will eventually deteriorate into "the crimson fire in which all sight burns."[20]

Colour Material

> *his prayers picked out the sunlight on the dead and silver sea and hardened it. It turned it into jewellery. The water was as solid as a silver plate. It rose from the distant valley into the mid-air haze.*
>
> *Quarantine (81)*

Crace's starving Galilean perceives the shimmering heat as water on the blistering sand. Like silver, the jewel-like feature of quartz,[21] the most abundant mineral in the Earth's crust, materialises the light into a solid heat haze. Unseen and unscattered, the long waved section of the electromagnetic spectrum then entangles the arid and soilless ecosystem into a heat trap.

The extreme temperature variation in deserts, as the rock and sand absorb heat during the day and release it at night, is matched by the force of winds and storms, due to the extremes of pressure variation, the mixing of cold and warm air and the erratic precipitation and evaporation of water. The violence of solar radiation in deserts is the source of an unforgiving intensity. They are crystalline places, and it is the gem-like quality of the sand grains that enhance its luminosity.

Thus the luminous desert flickers into being. The solar haze of heated air masses thickens the atmosphere into a corporality that is simultaneously entangled with and detached from the granularity of the quartz substrate. In Huxley's exploration of *Heaven and Hell* (1956), sand "was made up on a perfect geometrical pattern, with sharp angles, from each of which a brilliant shaft of light was reflected, while each tiny crystal shone like a rainbow."[22] Colour is decomposed light that intimates dispersion and obliquity of sight.

> *I stood and looked. The desert was outside my range, it was an alien being, it was science fiction, both saturating and remote, and I had to force myself to believe I was there.*
>
> *Point Omega (25)*

Against this crystal-rigidity, we also have the wavelike liquidity of light as it exists near absolute zero as photon condensation. This makes light into a droplet with surface tension that pours into our field of vision. The viscous droplets of perception gloop around the gaze that struggles to give form. It is the materiality of light and its constituent layering of wavelengths of electromagnetic radiation that is incarnate with vision and landscape. The luminosity of the landscape is expressed through the amplitude of light. The medium is indeed the message.

The fact that the crystal features centrally in physics as well as in philosophy and religion is not accidental. The state of all solid matter is essentially crystalline, that is, an orderly arrangement of particles that is nevertheless prone to become fluid. It facilitates the phenomenon of material growth as an expression of non-equilibrium. The crystal is also an ancient symbol of light and enlightenment because of its infinite tessellation. Huxley explores the luminous, "preternatural brilliance of colour" scattered through the crystalline structures of gemstones.[23] In the Ancient Buddhist notion of Indra's Net, the universe is conceived as an immense multidimensional lattice made up of jewels connecting together at intersecting nodes, each jewel refracting the light of every other jewel.[24] Such thinking references eschatological literature of all major religions in which "Heaven is always a place of gems."[25]

Material Vibration

> *the heat reverberated from the hot sand as from the slag of some molten sun, burning the rubber soles from his boots, and the light from the scattered flints of washed quartz flickered with diamond hardness. [...] the corona of glimmering phosphors which hung over the sand like myriads of fire-flies. [...] heliotropic jewels from the vermilion desert.*
>
> *The Cage of Sand (np)*

J. G. Ballard's desert is the final resting place of the fiery atoms of dead astronauts and satellites that circle the earth and re-enter the atmosphere in a "crimson fire". He re-iterates the purple shadows and vermillion glow that play across a transported desert, brought from Mars to balance out the Earth's spin, as actualised Deleuzian perception.

The scattering of light through crystal lattices narrates the desert as a coloured place, full of particles that diffract the white light and create "dense fields of colour completely independent of the form and real position of things".[26] Thought through the diffracting properties of quartz, and the reconfiguration of light patterns characterising a shift away from representational reflection, the human figure dissolves.

Vibration Dilution

> *a slow, painstaking figure, made thin and watery by the rising, mirage heat, as if someone had thrown a stone into the pool of air through which it walked and ripples had diluted it.*
>
> *Quarantine (13)*

Sight and perception are tempered in the desert. This is the first image of the young Galilean called Jesus, heading for quarantine in the desert, a "threatening conspiracy of rocks, wind and heat". Yet the final sight of his figure "no longer thin and watery, diluted by the mirage heat, distorted by the ripples in the air" (243) does not establish an epiphanic sudden revelatory instant of pure awareness. The perception of the novel's evil spirit, Musa, who has been resurrected by Jesus, and who now resurrects the Galilean in his own image, is false. It lacks, as Ponte describes it, the "total triumph" of pure visibility, the gaze that denatures form and its meaning and perceives "only the abstraction of light and colour."[27]

Dilution Colour

> *Behind him was translated light – blues and other colours shivering in the haze and sand.*
>
> *The English Patient (10)*

In Michael Ondaatje's story, the "English" patient also sees a vision, a mirage, this time of an archangel. The body of the healer appears enveloped in a curtain of glass bottles strung in line, full of ointments warmed by the sun and refracting the light. Light simulates sight as the man's perception penetrates the desert and dissects it into an abstraction in which "the world of tangibles and meanings dissolves into pure form and color."[28] The light acquires once again weight and pressures, density and grain.

Light radiates across the electromagnetic spectrum. The photonic energy stream that moves wave-like across the universe scintillates and interacts with the crystalline superheated structures of the desert sands, refracting in the shimmering illusion of the mirage bending upwards. The glittering desert lies under a clear sky. Free of aerosols, the vista is stretched into inscrutable distance, as light undoes the aerial perspective of the desert. The "spare lines, lucid spaces […] abstract austerities"[29] make the desert into lines of flight.

The material luminosity of light is understood though the substantial crystalline refraction. It illustrates the unison of idea and matter, perception and understanding. Light is not a torch that is shone unto material, making it visible in its distinct form, but instead light *is* form itself.

Interlude

> Each grain of sand, each bit of soil is diffracted/entangled across spacetime.[30]

The readings so far have illustrated that the desert is no longer in need of representation. It is already present. Light incarnates itself in the crystal structure of desert sand and arranges the desert as a photonic resonance captured on light-sensitive film. The desert as luminous landscape, as translucent space, exerts a powerful metaphorical pull on the imagination, but at the same time the desert light is spatial and material. Light, space and grain converge in the idea of diffraction.

Diffraction is reading the world though movement of light across spacetime, revealing the nature of matter through the potentiality of energy in patterns of interference. The identity of crystals, for example, is made visible through diffracting the extremely short wavelength light of X-rays. The diffraction patterns of desert quartz in particular are spectacular, showing vast numbers of reflections that reveal the complexity of perspectives.[31]

The symmetry of these diffraction patterns radiates from the axis of the light-beam in petal formations.[32] The scattering of the light through the multiple angles of the crystal reveals the diffraction fringes – darkness of light, the lightness of dark, the halo that diffuses the borders between, yet does not advocate indistinguishable similarity. Deleuze spoke of the lightning bolt, and its "dark precursor"[33] that references the dark within the light within the dark, that speaks to the inherent tension inside light and illumination – the subtle intimation of possibility and knowledge.

The diffracted light of a body detaches itself from the source and forms a corona and engenders multiple becomings.[34] Diffraction undermines the idea that light has a representative function of distinct entities. Even though diffraction is used to identify the crystal identity of matter, the overlapping diffraction cones defy specificity. Also, a static image of the diffraction patterns belies their dynamic nature, as a pattern is undone as much as created in fractions, constantly and iteratively repatterned and "worlded" as "spacetimemattering," similar to the desert whose unchanging vistas are, in fact, constant velocities.

Let's move from the physical event of diffraction to diffraction as a methodological lens,[35] in which a linear process of reading is transmuted into a dialogic "reading through one another" spreading and bending meaning around words.[36] Such diffractive reading gives space to a creative response to material and narrative by considering the text as wave in which aspects of language and landscape intra-act. It supports a reading event in which place and text are not, in Edmond's words, "passive recipients of wave forms."[37] They are dynamic intra-actions that do not register as cause and effect, but as inherent relations that exist prior to but are also contingent with the reading event.[38] The storied desert can thus be read diffractively as wave.

> *in the desert it is easy to lose a sense of demarcation. When I came out of the air and crashed into the desert, into those troughs of yellow, all I kept thinking was, I must build a raft ... I must build a raft.*
>
> *The English Patient (20)*

Reading the desert, we experience it. The desert light does not show what is, but refracts what *isn't*. The yellow sands hold distant memories of water, "the ghost between your hands and your mouth" (20). It *isn't* the haunting of water that resides in aquifers deep down in the earth's crust and now plays as a shimmer on the desert surface. The canyons dip down a "thousand years" (21). It *isn't* a past reworked into a future, as the Bedouins "spilled and slid over sand and stones" (21), themselves a fluidity shifting "like a spray of milk or a wing" (21). The tribal memories of water-logged strata of desert are entangled with the future and thus "enfolded memory" becomes enfolded desire.[39]

> *I looked out into the blinding tides of light and sky and down toward the folded copper hills that I took to be the badlands, a series of pristine ridges rising from the desert floor in patterned alignment. Could someone be dead in there? I could not imagine this. It was too vast, it was not real, the symmetry of the furrows and juts, it crushed me, the heartbreaking beauty of it, the indifference of it, and the longer I stood and looked the more certain I was that we would never have an answer.*
>
> *Point Omega (116–117)*

Read through Ondaatje's haunted desert-ocean, DeLillo's palimpsest of dilated space and time rides high on the diffractive energy waves of the desert. The text's spatial form and filmic patterning reflects how waves of cognition, sound and image intra-act in the desert, which is no mere background but central holograph that transposes light into language, bending into different directions.[40]

> *the landscape began to seem normal, distance was normal, heat was weather and weather was heat. I began to understand what Elster meant when he said that time is blind here. Beyond the local shrubs and cactus, only waves of space, occasional far thunder, the wait for rain, the gaze across the hills to a mountain range that was there yesterday, lost today in lifeless skies.*
>
> *Point Omega (81)*

Read through DeLillo's cinematic language, the long view now dissolves these waves of space:

> *As he reads, we dissolve to a slope among the foothills of the San Gabriel mountains. Naked but for a cactus here and there, the stony ground lies dead and mangled in the sunshine. A network of ramifying gullies furrows the hillside. Some of them are still in the infancy of erosion, others have cut their way deep into the ground.*
>
> *Ape and Essence (116)*

Read though William Tallis's screenplay, the "English" patient's journal picks up the dissolution. "Erase the family name!" (148). The landscape erases history, nationhood and civilisation:

> *Up close the glass was rough and sandblasted, glass that had lost its civilisation.*
>
> *The English patient (10)*

Read though Ondaatje's elegy of desert winds that obliviate armies, Miri's night spent in the desert grave she dug for her husband Musa is full of "discord and detonation":

> *Earth collapses with the engineering of the ants; lizards smack the pebbles with their tails, the sun fires seeds in salvos from their pods.*
>
> *Quarantine (14)*

Diffraction challenges the notion of deficit. Desert as absence of life, naked, under a lifeless, barbaric sky, is queered to desert as abundance of light, firing salvos. The combined effect of light and life in the desert is evident in the motion that belong to each, a motion that radiates and overlaps, creating sound- and lightscapes of oscillating waves.

As the photon is both particle and wave, so the desert is both ontologically energy and matter. Both are contingent. This relational "*difference within*" is an event, not a state. It happens as an entangling, or as Barad puts it, a "*cutting together-apart*". "Each bit of matter, each moment of time, each position in space is a multiplicity, a superposition/entanglement of (seemingly) disparate parts".[41] This superposition is clearly evident in the indeterminate and unintelligible desert landscape.

Referencing Heisenberg's Uncertainty Principle, we create what we interpret. Our diffraction apparatus (text, reader, reading, perceiving) creates the desert at condensed nodes in the changing diffracted field that is forever repatterned. Desert is storied matter, "sedimented intra-acting, an open field,"[42] an agentive and productive potential, not a spatial given. Kaiser cites Èdouard Glissant: "we no longer *describe* landscapes, we speak and breathe them."[43] Rather than judging the text, we are involved in the (re-)production of it.

Divi sūrya-sahasrasya – 1000 suns in the sky[44]

The perception of deserts as uninhabitable wastelands belies the "sublimity of space" that is filled with a "Baroque extravagance" of "the queer, uncanny vegetable shapes."[45] Such ecological richness is commonly unobserved by human visitors. D. H. Lawrence, for example, wrote of the somnambulists who ride through the desert asleep and unconscious of the "ash-white hearth of the desert".[46] This ash-white heart nevertheless now contains stores of illegally dumped waste, which aids the signification of the desert as wasteland.[47] Waste also informs the (ab)use of the desert as Nuclear Test Site, mining operation or militarised zone. Considered to be the "most profoundly devastated area of the nuclear landscape,"[48] the Nevada National Security Site (formerly the Nevada Test Site) illustrates the high density of energy in all forms. It speaks to the (re-)production of desert as luminous landscape "in which all sight burns."[49]

The atmospheric atomic detonations in the Nevada desert in the 1950s turned the luminous desert into a different sort of "Baroque extravagance," as the elemental force of nuclear fission expanded the horizon and ignited the sand into a crystalline sea of light. The morphology of the landscape was shaped by and shaped itself the vegetable form of the explosion. The "unearthly light" that is reported at the first test explosions is linked to a creation story in which the "blinding light seems born directly from the darkness and nothingness of the desert."[50] The verticality of nuclear explosions, in which the earth is boiled and churned in an upwards draft, crowned by the furious light of the eruption, is heightened by the extensive horizontality of the desert's smooth sandy layers.

> *Far off, from below the horizon, a column of rosy smoke pushes up into the sky, swells out into the likeness of an enormous toadstool and hangs there, eclipsing the solitary planet.*
>
> *Ape and Essence (34–35)*

The pillar of fire that haunted the Israelites in the desert now haunts humanity as a whole through its linkage of sky and earth, past and future.

Thermonuclear Time

> *the series of weapons tests had fused the sand in layers, and the pseudogeological strata condensed the brief epochs, microseconds in duration, of thermonuclear time.*
>
> *The Terminal Beach (np)*

In J. G. Ballard's post-atomic story, the sand of the terminal beach refracts via "crystallizing the immense epoch of time." The desert is a bare geology in which the bones of the earth are revealed through "processes of erosion and calcification"[51] The physical requirement of an "empty" and "barren" landscape was deemed important to absorb the energy and radiation and give space to the spectacle of the nuclear fireball that rose from these flats. The desert conditions with its clear atmosphere and flat terrain are all factors in determining the luminance of the explosion.[52]

The Soviet nuclear tests were conducted on Novaya Zemlya (New Land), an island finger reaching into the Arctic, dividing the Barents from the Kara Sea. The location of perhaps the largest atmospheric detonation of a nuclear bomb is a different type of desert, a permafrosted arctic plane. Now part of the Arctic desert ecoregion, Novaya Zemlya gives its name, appropriately enough, to the Novaya Zemlya effect,[53] a polar mirage in which sunlight refracts between the inverted thermal layers to give the impression of a "rectangular sun."

Time Sunrise

Atomic tests wrote the iconography of nuclear light. The very first atomic test, known as "Trinity," detonated on 16 July 1945 in the New Mexico desert and created a visual language that focused on the contrast between the white light and black earth and sky against which it was set. Berlyn Brixner's iconic photograph shows the nuclear blast as a sunrise, with the dome of the explosion an intact sphere as it rises from the horizon, encircled by a skirt of superheated air and sand.[54] The incandescent surface of the fireball also resembles a jellyfish in its pellucid quality of the black and white photograph. The violent turbulence inside the gas cloud turns the superheated air opaque.[55]

The monochrome aspect of atomic photography is striking, thriving on the contrast between the shadow of absences (as in, e.g., the carbonised shadows of victims in Hiroshima photographed by Yoshito Matsushige) and the "unearthly" lightflash. Catching the light of a nuclear sun on photonic plates and film, the eternally present visual information merged with the layered past, as light travels from the past towards the immeasurable future in an apocalyptic instant.

The haptic capture of the image via the camera lens[56] allows for some sensory distance to the enormity of the light. But, as Deleuze stipulates: "Visibilities are not forms of objects [...] but forms of luminosity which are created by the light itself and allow a thing or object to exist only as a flash, sparkle or shimmer."[57] As in the earlier contemplation of the cinematic desert, the chemical trace of photons on film, like the electro-chemical irritation of neurons via the optic nerve, acts on medial substrates that capture forms of luminosity.

As Bryan C. Taylor points out, the inherent association of radiation and photography is one of wavelength.[58] The longer waves of visible light interact with the light-sensitive chemical layer on, for example, celluloid. The shorter X-ray waves are used diagnostically in diffraction tests, but more widely, of course, as part of medical investigations. The yet shorter and violently penetrative beta- and gamma-rays, the highest-energy form of light that emerges from dying stars and supermassive black holes, exists on earth in the natural short-lived emanation of the lightning bolt,[59] but more insidiously in the decay of radioactive elements.

Sunrise Decay

> *I perceived that the whole desert was alive with the faint, flickering glow of some phosphorescent decay.*
>
> *"My quickened senses told me that this interpretation of the pale, greenish light was indeed correct: the desert, the monuments, nay, the whole planet, were passing through their dissolution to an everlasting nothingness. For the phosphorescent flickerings were the evidence that the atoms of the metals and the sands of the Desert were disintegrating into their primordial electricities. The incessant ejections of those minute particles of electricity rendered visible the dying matter—the metals of the monuments and the sands of the Desert—of which they were the natural, inevitable decay." [...] For I saw the infinitely slow natural breaking down of matter into energy as one continuous stream of flickering light.*
>
> *The Time Stream (np)*

John Taine's *The Time Stream* extols the otherworldliness of the desert through the trope of a desert planet in which the desert is decaying matter – not due to an atomic explosion, but simply via the inexorable decay of entropy.

Decay of matter is a decay into colour is a diffraction of light. As matter disintegrates it becomes colourful, patina, rust, Verdigris. As matter vapourises, its energy increases exponentially and becomes pure light. The radiant energy of an atomic blast, light that "exceeds all other man-made sources of light", as Clay P. Butler describes it in factual terms, covers the whole range of the spectrum, with the bright white flash covering around 50 percent of the thermal radiation that was created at point of fission. He describes the sequence of colours created in the explosion, "from green to red" as "awesome."[60] Awe, derived from "terror" in Old English, re-codes the sublimity of the desert into one of "terrific" spectacle.

The holography of the lightning blast is a concrete abstraction of the luminous desert. In a holograph, the light-field of atomic lattices reorganises itself by superimposing deadly short-waves onto the light, creating a brilliant flash and an array of coloured lights.[61] This extension of luminosity in terms of the "unearthly light" of the first atomic explosions in the 1940s, ties the desert to a dynamism that is integral to "spacetimemattering." The "dark precursor" of the lightning bolt as a natural source of gamma radiation finds its twin in the violet haze of the explosion that contains the dark within the light within the dark. The aesthetic effect of the light of a thousand suns de-anthropomorphises the landscape to the extent that it becomes pure intensity.

But like the crystal light at the beginning, the "inconceivable light" of the atomic sublime was a diffracted swirl of colour rather than a simple blinding white light. Like the coloured desert, the afterglow of radiant thermal intensity left a crimson band across the vacant horizon to engender perception. Eyewitnesses recalled bluish-green light, a "pillar of purple fire", a rose-coloured flower petal.[62] Thomas Farrell, who witnessed Trinity, spoke of a searing light that "was golden, purple, violet, gray and blue."[63]

Hence, the desert carries its signification further as extreme materiality. Every grain of sand pulverised in the atomic explosion and distributed as crystalline and glassy fallout debris disrupts perspective and space through intensities of colour. The high-frequency radiation of gamma rays, violent eruptions of the electromagnetic spectrum of photonic energy, leaves traces in all matter. It irradiates genes, as well as photographic film, and thus the invisible becomes visible.[64] The desert at time of explosion becomes a negative of itself, an oblivion. These non-traces under the clear skies of the desert are scattered across vast distances, as the auroreal displays of nuclear explosions can extend beyond horizons. But the light contracts the openness of space around the single blast point, dissected by the sheer geometry of the desert horizon.

> *there is a way a small cargo plane will come down to land, slipping from the level of horizon. It tips its wings within desert light and then sound stops, it drifts to earth. [...] They dip their wings across the light and then they enter that silence.*
>
> *The English Patient (272)*

The plane crashes. The form intra-acts with light intra-acts with silence intra-acts with the fire-ball of the crashed plane. Yet, the luminous desert is pure space that swallows light, sound and time as the quartz matter of the desert vibrates unendingly at a precise frequency, transmitting pips to a non-listening aether. The visionary quality of intense light mingles with the spiritual meaning of absence, the "sheer pure absolute," the not-it of *spacetimematter*.

The desert is the plane of immanence, "a section of chaos that acts like a sieve."[65] The unactualised ideas on the plane are sifted into possible becomings through encountering forces and counterforces. The desert is thus pure potentiality and photonic energy, strumming a rhythm that makes the chaos into becomings. Its sand runs through the sieve, everlasting, never-ending, under a translucent sky.

Notes

1 Karen Barad, *Meeting the Universe Halfway: Quantum Physics and the Entanglement of Matter and Meaning* (Duke University Press, 2016). Cf. an "ontological performance of the world in its ongoing articulation" (149).
2 Karen Barad, "Diffracting Diffraction: Cutting Together-Apart," *Parallax* 20, no. 3 (2014): 169, https://doi.org/10.1080/13534645.2014.927623.
3 Cf. Aldous Huxley, *Beyond the Mexique Bay* (Chatto and Windus, 1950), 240.
4 Gilles Deleuze and Felix Guattari, *What Is Philosophy?*, transl. Hugh Tomlinson and Graham Burchell III (Columbia University Press, 1994), 50–51.
5 Gilles Deleuze, *Essays Critical and Clinical*, transl. D. W. Smith and M. A Greco (University of Minneapolis Press, 1997), 116.
6 Deleuze and Guattari, *What Is Philosophy?, 36.*
7 Hanjo Berressem, *Gilles Deleuze's Luminous Philosophy* (Edinburgh University Press, 2020), 170.
8 Deleuze and Guattari, *What Is Philosophy?*, 49.
9 Berressem, *Luminous Philosophy*, 171–172.
10 Deleuze, *Essays Critical and Clinical*, 115.
11 Deleuze, *Essays Critical and Clinical*, 115–116.
12 Aldous Huxley to Julian Huxley, October 27, 1946, in *Letters of Aldous Huxley*, ed. Grover Smith (Chatto and Windus, 1969), 554.
13 Aldous Huxley, "The Doors of Perception," in *Complete Essays*, Volume V, eds. Robert S. Baker and James Sexton (Ivan R, Dee, 2002), 178–180.
14 Huxley, *Doors of Perception*, 178–180. Deleuze, *Essays Critical and Clinical*, 115.
15 Huxley quotes from the Bardo Thodol. Aldous Huxley, "Ozymandias," in *Complete Essays*, Volume V, eds. Robert S. Baker and James Sexton (Ivan R, Dee, 2002), 311–312.
16 Berressem, *Luminous Philosophy*, 2.
17 Berressem, *Luminous Philosophy*, 213.
18 Gilles Deleuze, *Cinema 2: The Time Image*, transl. Hugh Tomlinson and Robert Galeta (Bloomsbury, 2013), 84.
19 Deleuze, *Essays Critical and Clinical*, 116.
20 Deleuze, *Essays Critical and Clinical*, 116.
21 It comes in a variety of vitreous glassy lustres that straddle the range of rainbow colours of visible light. For images of quartz crystals and other sand, see e.g. https://geology.com/stories/13/sand/ as well as Gary Greenberg's photographs at http://sandgrains.com/
22 Aldous Huxley, "Heaven and Hell," in *Complete Essays* Volume VI, eds. Robert S. Baker and James Sexton (Ivan R. Dee, 2002), 10.
23 Huxley, "Heaven and Hell," 14.
24 Leslie Paul Thiele, *Indra's Net and the Midas Touch: Living Sustainably in a Connected World* (MIT Press, 2011), 18.
25 Huxley, "Heaven and Hell," 13.
26 Alessandra Ponte, "The House of Light and Entropy: Inhabiting the American Desert," *Assemblage*, No. 30 (August 1996): 21.
27 Ponte, "The House of Light and Entropy," 21.
28 Ponte, "The House of Light and Entropy," 20.
29 Ponte, "The House of Light and Entropy," 14.
30 Barad, "Diffracting Diffraction," 184.
31 Xian-Rong Huang et al., "Correct Interpretation of Diffraction Properties of Quartz Crystals for X-ray Optics Applications," *Journal of Applied Crystallography* 51, no. 1 (2018): 146, https://doi.org/10.1107/S1600576717018155.
32 See e.g. "Laue Diffraction," PhysicsOpenLab: Modern DIY Physics Laboratory for Physics Enthusiasts, nd, accessed August 7, 2023, https://physicsopenlab.org/2018/01/18/laue-diffraction/.
33 Berressem, *Luminous Philosophy*, 5.

34 Berressem, *Luminous Philosophy*, 8.
35 Barad, *Meeting the Universe Halfway*, 89–90.
36 Barad, *Meeting the Universe Halfway*, 30.
37 Jacob Edmond, "Diffracted Waves and World Literature," *Parallax* 20, no. 3 (2014): 247, https://doi.org/10.1080/13534645.2014.927632.
38 Cf. Derek Attridge, *The Singularity of Literature* (Routledge, 2017).
39 Karen Barad, "Quantum Entanglements and Hauntological Relations of Inheritance: Dis/continuities, Spacetime Enfoldings, and Justice-to-come," *Derrida Today* 3, no. 2 (2010): 261.
40 John Cayley, "Beyond Codexspace: Potentialities of Literary Cybertext," *Visible language* 30, no. 2 (1996): 164.
41 Barad, "Diffracting Diffraction," 175–176.
42 Barad, "Diffracting Diffraction," 169.
43 Birgit Mara Kaiser, "Worlding CompLit: Diffractive Reading with Barad, Glissant and Nancy," *Parallax* 20, no. 3 (2014): 284, https://doi.org/10.1080/13534645.2014.927634.
44 Swami Mukundananda, *Bhagavad Gita, The Song of God*, Chapter 11, 2014, https://www.holy-bhagavad-gita.org/chapter/11/verse/12.
45 Aldous Huxley, "Unpainted Landscapes," in *Complete Essays* Volume. VI, eds. Robert S. Baker and James Sexton (Ivan R. Dee, 2002), 346.
46 D. H. Lawrence, "Men in New Mexico," in *The Complete Poems of D.H. Lawrence* (Wordsworth Poetry Library, 1994), 335–336.
47 Cf. the recent discovery of vast dumps of discarded clothes in the Atacama desert. Cf. https://www.ecowatch.com/chile-desert-fast-fashion-2655551898.html.
48 Valerie L. Kuletz, *The Tainted Desert: Environmental and Social Ruin in the American West* (Routledge, 2016), 70.
49 Deleuze, *Essays Critical and Clinical*, 116.
50 Emily Eliza Scott, "Birth of the Atomic Desert," (PhD diss., University of California, 2003), 7.
51 Ponte, "The House of Light and Entropy," 23.
52 This in turn ignited, according to Kirsch, the imagination of the people. Scott Kirsch, "Watching the Bombs Go Off: Photography, Nuclear Landscapes, and Spectator Democracy," *Antipode* 29, no. 3 (1997): 227–255.
53 See e.g. Siebren van der Werf et al., "Novaya Zemlya Effect and Sunsets," *Applied Optics* 42, no. 3 (2003): 367–378.
54 "Trinity," The Nuclear Weapon Archive: A Guide to Nuclear Weapons, nd, accessed August 7, 2023, https://nuclearweaponarchive.org/Usa/Tests/Trinity.html
55 Cf. Deleuze, Essays Critical and Clinical, "luminosity itself becomes opaque", 115.
56 Cf. "fingery eyes," as used in Donna Haraway, *When Species Meet*. Vol. 3 (University of Minnesota Press, 2008), 5, http://xenopraxis.net/readings/haraway_species.pdf
57 Gilles Deleuze, *Foucault*, quoted in Berressem, *Luminous Philosophy*, 187.
58 Bryan C. Taylor, "Nuclear Pictures and Metapictures," *American Literary History* 9, no. 3 (1997): 588.
59 Gregory S. Bowers et al., "Gamma Ray Signatures of Neutrons from a Terrestrial Gamma Ray Flash," *Geophysical Research Letters* 44, no. 19 (2017): 10,063–10, 070.
60 Clay P. Butler, "The Light of the Atom Bomb: In Brightness, a Nuclear Detonation is Comparable to the Sun," *Science* 138, no. 3539 (1962): 483, 489.
61 Butler, "The Light of the Atom Bomb," 483–489.
62 Peter Bacon Hales, *Outside the Gates of Eden: The Dream of America from Hiroshima to Now* (University of Chicago Press, 2014), 11. Hales quotes William L Laurence's record of the bombing of Nagasaki, p. 15. In fact, the name for the explosion, rather than a mushroom, was 'rose'.
63 Quoted in Peggy Rosenthal, "The Nuclear Mushroom Cloud as Cultural Image," *American Literary History* 3, no. 1 (1991): 80.

64 James H. Webb, "The Fogging of Photographic Film by Radioactive Contaminants in Cardboard Packaging Materials," *Physical Review* 76, no. 3 (1949): 375–380.
65 Deleuze and Guattari, *What Is Philosophy?*, 42.

Works Cited

Attridge, Derek. *The Singularity of Literature*. Routledge, 2017.

Ballard, James G. "The Cage of Sand." *New Worlds*, no. 119. 1962. Accessed April 24, 2023. https://readerslibrary.org/wp-content/uploads/The-Cage-of-Sand.pdf.

Ballard, James G. "The Terminal Beach." *New Worlds*, no. 140. 1964. Accessed April 24, 2023. https://horrorworkout.files.wordpress.com/2013/09/jg_ballard_terminal_beach.pdf

Barad, Karen. "Diffracting Diffraction: Cutting Together-Apart." *Parallax* 20, no. 3 (2014): 168–187. https://doi.org/10.1080/13534645.2014.927623.

Barad, Karen. *Meeting the Universe Halfway: Quantum Physics and the Entanglement of Matter and Meaning*. Durham: Duke University Press, 2016.

Barad, Karen. "Quantum Entanglements and Hauntological Relations of Inheritance: Dis/Continuities, Spacetime Enfoldings, and Justice-to-Come." *Derrida Today* 3, no. 2 (2010): 240–268.

Berressem, Hanjo. *Gilles Deleuze's Luminous Philosophy*. Edinburgh University Press, 2020.

Bowers, Gregory S., David Miles Smith, G. F. Martinez-McKinney, M. Kamogawa, S. A. Cummer, J. R. Dwyer, D. Wang, M. Stock, and Z. Kawasaki. "Gamma Ray Signatures of Neutrons from a Terrestrial Gamma Ray Flash." *Geophysical Research Letters* 44, no. 19 (2017): 10063–10070.

Butler, Clay P. "The Light of the Atom Bomb: In Brightness, a Nuclear Detonation Is Comparable to the Sun." *Science* 138, no. 3539 (1962): 483–489.

Cayley, John. "Beyond Codexspace: Potentialities of Literary Cybertext." *Visible language* 30, no. 3 (1996): 164–183.

Crace, Jim. *Quarantine*. Penguin, 1998 (1997).

Deleuze, Gilles. *Cinema 2: The Time Image*. Translated by Hugh Tomlinson and Robert Galeta. Bloomsbury, 2013.

———. *Essays Critical and Clinical*. Translated by Daniel W. Smith and Michael A. Greco. Minneapolis: University of Minneapolis Press, 1997.

Deleuze, Gilles and Felix Guattari. *What Is Philosophy?* Translated by Hugh Tomlinson and Graham Burchell III. Columbia University Press, 1994.

DeLillo, Don. *Omega Point*. Picador, 2011 (2010).

Edmond, Jacob. "Diffracted Waves and World Literature." *Parallax* 20, no. 3 (2014): 245–257. https://doi.org/10.1080/13534645.2014.927632.

Hales, Peter Bacon. *Outside the Gates of Eden: The Dream of America from Hiroshima to Now*. University of Chicago Press, 2014.

Haraway, Donna. *When Species Meet*, Volume 3. University of Minnesota Press, 2008. http://xenopraxis.net/readings/haraway_species.pdf

Huang, Xian-Rong, Thomas Gog, Jungho Kim, Elina Kasman, Ayman H. Said, Diego M. Casa, Michael Wieczorek, Marcelo G. Hönnicke, and Lahsen Assoufid. "Correct Interpretation of Diffraction Properties of Quartz Crystals for X-Ray Optics Applications." *Journal of Applied Crystallography* 51, no. 1 (2018): 140–147. https://doi.org/10.1107/S1600576717018155.

Huxley, Aldous. *Ape an Essence*. Flamingo, 1994 (1949).

———. *Beyond the Mexique Bay*. Chatto and Windus, 1950.
———. "Heaven and Hell." In *Complete Essays*, Volume VI, edited by Robert S. Baker and James Sexton, 5–49. Ivan R. Dee, 2002.
———. "Letter to Julian Huxley, October 27, 1946." In *Letters of Aldous Huxley*, edited by Grover Smith, 551–555. Chatto and Windus, 1969.
———. "Ozymandias." In *Complete Essays*, Volume V, edited by Robert S. Baker and James Sexton, 301–212. Ivan R, Dee, 2002.
———. "The Doors of Perception." In *Complete Essays*, Volume V, edited by Robert S. Baker and James Sexton, 157–191. Ivan R, Dee, 2002.
———. "Unpainted Landscapes." In *Complete Essays*, Volume VI, edited by Robert S. Baker and James Sexton, 340–347. Ivan R. Dee, 2002.
Kaiser, Birgit Mara. "Worlding CompLit: Diffractive Reading with Barad, Glissant and Nancy." *Parallax* 20, no. 3 (2014): 274–287. https://doi.org/10.1080/13534645.2014.927634.
Kirsch, Scott. "Watching the Bombs Go Off: Photography, Nuclear Landscapes, and Spectator Democracy." *Antipode* 29, no. 3 (1997): 227–255.
Kuletz, Valerie L. *The Tainted Desert: Environmental and Social Ruin in the American West*. Routledge, 2016.
Lawrence, D. H. "Men in New Mexico." In *The Complete Poems of D.H. Lawrence*, 335–336. With an introduction and notes by David Ellis. Ware: Wordsworth Poetry Library, 1994.
Ondaatje, Michael. *The English Patient*. Bloomsbury, 2004 (1992).
PhysicsOpenLab: Modern DIY Physics Laboratory for Physics Enthusiasts. "Laue Diffraction." Accessed August 7, 2023. https://physicsopenlab.org/2018/01/18/laue-diffraction/.
Ponte, Alessandra. "The House of Light and Entropy: Inhabiting the American Desert." *Assemblage*, no. 30 (August 1996): 12–31.
Rosenthal, Peggy. "The Nuclear Mushroom Cloud as Cultural Image." *American Literary History* 3, no. 1 (1991): 63–92.
Scott, Emily Eliza. "Birth of the Atomic Desert." PhD diss., University of California, 2003.
Swami Mukundananda. *Bhagavad Gita, The Song of God*. 2014. Bhagavad Gita, The Song of God – Swami Mukundananda (holy-bhagavad-gita.org).
Taine, John. *The Time Stream*. 1931. Accessed April 24, 2023. https://www.gutenberg.ca/ebooks/tainej-timestream/tainej-timestream-00-h.html.
Taylor, Bryan C. "Nuclear Pictures and Metapictures." *American Literary History* 9, no. 3 (1997): 567—597.
The Nuclear Weapon Archive: A Guide to Nuclear Weapons. "Trinity." Accessed August 7, 2023. https://nuclearweaponarchive.org/Usa/Tests/Trinity.html
Thiele, Leslie Paul. *Indra's Net and the Midas Touch: Living Sustainably in a Connected World*. MIT Press, 2011.
van der Werf, Siebren Y., Günther P. Können, and Waldemar H. Lehn. "Novaya Zemlya Effect and Sunsets." *Applied Optics* 42, no. 3 (2003): 367—378.
Webb, James H. "The Fogging of Photographic Film by Radioactive Contaminants in Cardboard Packaging Materials." *Physical Review* 76, no. 3 (1949): 375–380.

PART 2

Poetics of Aridity

5

REVISITING *PALAI*

Imaginings of Arid Lands in Tamil Literary and Cultural Texts

Chitra Sankaran and Gayatri Thanu Pillai

Introduction: *Thinai* and *Palai*

Literary and cultural conceptualizations of the desert have been complex, varied, and deeply rooted in contextual and sociocultural specificities. However, existentialist abstractions have dominated the discourse on deserts. From wilderness to wasteland, desert topologies have been imbued with moral, philosophical and aesthetic correlations. Literary scholar Stephen Bending observes that "landscapes are created by omission and suppression, both visual and ideological…indeed, a physical viewpoint is itself likely to be determined in turn by ideological assumptions about class and aesthetics, beauty and utility, politics and power."[1] Western metaphysics, particularly as reflected in literary and cultural texts, often projects the desert as a confrontational space between fragility and mortality. As the discourse on wasteland aesthetics widens, it is imperative to reevaluate some of the core assumptions circulating on the subject while also foregrounding how literature and culture from the Global South have articulated anxieties using deserts and arid wastelands.

It is a widely acknowledged fact, and one deplored by ecolinguists, that ecological discourses of dominant cultures still adopt an anthropocentric lens in approaching nature. Martin Doring and Francesca Zunino[2] observe that though binaries such as "nature and culture" should be obsolete, they continue to proliferate in dominant global discourses, while lexical items that sharply divide the human from the environment still prevail. Unfortunately, such renderings marginalize or obliterate other language systems that assert alternative world views that are more compatible with contemporary scientific perspectives, which underscore humans' embedment within their ecospheres.

Highlighting the importance of reframing and revising conventional ontological categories that have largely focused on the Global North, this chapter presents

DOI: 10.4324/9781003397212-8

important non-western perspectives on wasteland aesthetics from Tamil literature (both Indian and diasporic). As Aidan Tynan points out, "it is not at all obvious where the distinction between the real and rhetorically constructed desert lies. In any case, all places are, to one extent or another, rhetorical or semiotic constructions."[3] With this as premise, we introduce and explore a unique ecological concept, *Thinai*, from ancient Tamil, that was conceptualized during its *Sangam* period (2nd century BCE to 2nd century CE).[4] *Thinai* is based on the understanding of humans as embedded in their ecological habitats and, therefore, conjoins the traditional binaries of nature and culture. Focusing on rhetorical strategies such as the use of imagery and literary allusions, we then go on to discuss how one of the five *thinai*[s], *palai*, which refers to dry or arid land, is inscribed within the *thinai* system in ancient Sangam poetry to denote certain landscape associations that are markedly different from those in western cultural imagination. Finally, we analyze productive, contemporary revisitations of the *palai thinai* in fiction, poetry and the visual media.

Thinai realigns the Nature/Culture binary in compelling ways. The five *thinais* (*aiynthinai*) are the five types of landscapes found in *Tamilagam*.[5] These are *kuṟiñci* (mountainous terrain), *mullai* (forested land), *marutham* (agricultural land), *neythal* (coastal land), and *palai* (dry land).[6] These regions have been rigorously encoded with aesthetic theories that prescribe poetic formulas and usage. For example, through each of these *thinais, akam* (interior) poetry dealing with love, and *puram* (exterior) poetry dealing with themes of war engage with three distinct sets of rules. The first is known as *mutal*[7] that relates to the nature of the land, the season or time that corresponds with each *thinai*. The second, *karu*[8] prescribes the seed theme that presents the relationship between human beings, their immediate society or the world at large. This delineates the flora, fauna, musical instruments, gods and goddesses associated with each *thinai*. The last, *uri*[9] details the various phases of love, war and distinctive moods associated with each *thinai*. The rules of *uri, mutal* and *karu* determine the mood, subject, characters and setting of the poem. Thus, as a gestalt, *thinai* prescribes and circumscribes the corresponding region's characteristic features and the culture and the conduct of its inhabitants. It operates as a composite and all-encompassing concept that goes beyond landscape.

What *thinai* emphasizes is that human activities in *akam* and *puram* are inseparable from the land in which they are embedded. The rules of *thinai,* which make it not only unethical but against accepted poetic norms to write about emotion independent of the land, highlight the deeply felt ethic that inner and outer boundaries are artificial. Like Haiku, *thinai,* as codifier, is able to paint a vivid picture of a whole landscape and its community with just a few symbols. For instance, the mention of a plant or an occupation can signal the *thinai*. Thus, poetry in *kurinci thinai,* set in mountainous regions, sings of the union of lovers; *mullai thinai* uses the forest as its setting, with "waiting" as its theme; *marutham,* connected with agricultural land, concerns itself with quarrels between lovers, while *neythal* or coastal land is associated with pining. Finally, *palai* which signifies dry, arid wasteland or desert is themed around separation of lovers.

Described as a dry terrain that emerges from the withering of other landscapes under the heat of the blazing sun, land in *palai thinai* is often depicted as inhospitable and inhabited by bandits and highwaymen. The concept of *thinai* was so deeply integrated with classical Tamil grammar and rules for poetry such that it would have been 'ungrammatical' to sing of lovers' joy and union in arid land or *palai thinai*. Such transgressions would have constituted erroneous poetry.[10] In the ensuing textual discussions, we present some examples of classical Sangam poetry that exemplify *palai thinai* and discuss some contemporary revisitations of it to show how *Thinai* has been reworked to facilitate a discourse that negotiates and challenges oppressive, hegemonic structures, binaristic value-dualisms and grand narratives that glorify patriarchy and war. In these contemporary texts, we see innovative and powerful realignments of the prescribed codes of *palai thinai* that reimagine spaces and the emotions associated with arid landscapes in provocative ways.

Palai in Sangam Poetry

In Sangam poems written in *palai thinai*, dangerous journeys are undertaken by the hero through desert areas, often in search of livelihood. The hardships induced by the inhospitable terrain are vividly described. The time of the day associated with these parched wastelands is noon and the season is the scorching summer. Wild animals roam unfettered, and cacti are the only vegetation. Bandits add to the treacherous and perilous nature of the landscape.

Tholkappiyam, the most ancient extant Tamil compendium, comments comprehensively on orthography, etymology, morphology, linguistic rules, prosody, and the significance of context in Tamil language and literature by citing examples from texts that presumably pre-dated it and many of which are now no longer known. Poems from *Kurunthokai*, an anthology of about 400 poems, reveal how *palai thinai* determines the codified setting and the theme, discussed above. An example is the poem numbered 174:

> In the desolate, rain-forsaken land
> the twisted *kalli*'s [cactus's] pods
> open with a crackle
> frightening the mating pigeons
> with their dense, downy feathers.
>
> He has left me languishing
> avowing to find wealth.
> not minding the manifold risks in his path.
>
> It needs to be acknowledged
> that in this world
> wealth has all the support
> while love must stand in the margins.[11]

In this poem, we see the *palai* theme of separation is centered. The reference to the lack of rain and the *kalli* or cactus pods signifies the desert landscape in which the poem is set. Adhering to the *thinai* conventions of separation as theme for *palai thinai*, the speaking voice in this poem is a woman whose lover has abandoned her, and she is left pining for him. In another *Kurunthokai* poem (number 20), the distraught and abandoned female lover declares passionately:

If it be strength
to shake off love and tenderness
and depart in pursuit of wealth
leaving us to languish,
then, let him be strong.
And let us womenfolk be
fools in love.

Both of these ancient poems, in accordance with *palai thinai,* retain the theme of separation, underscoring the aridity of both landscape and mindscape. These women are vocal, assert their emotional bereavement, and are passionate in their expression of loss. This is even more evident in *Kuruthokai* 27 by Velliveethiyar where the bold assertion of her sexual needs by the female poetic voice contravenes the strong social taboo that exists in later centuries right up to the current millenia, that in mainstream Tamil literature at least, inhibits such bold sexual assertions by women.

Like the delicious milk of a cow
that is spilt on the ground
goes to waste
without being suckled by the calf
or collected in a vessel,
so too does the sparkling brightness of my complexion,
eaten away
by the pallor of lovesickness;
enjoyed neither by my lover
nor relished by me.

Finally, in *Kurunthokai* 124, the cruelty of separation is foregrounded in the female lover's friend's parting speech to the departing lover:

You say that the wilderness
with bare *Omai* [Mango] trees
that bespeak the desolation
of a deserted village

and the meandering path
of the salt mongers
passing by in groups
would be an awful place
to take her.
But try convincing her
that home would be
sweet to her in her loneliness.

Though *thinai* as a literary regulatory concept has long since vanished in Tamil literature, the idea of the five *thinai*[s], and especially, their joyous or dark imaginings have had a subtle influence over Tamil writers despite its waning influence through the centuries. The themes from the *palai thinai* discussed above, have been effectively utilized by contemporary poets, speaking not only through *akam* poems themed around love but also through *puram* poems themed around war. A prime example of this can be seen in the poem, "*Yarukkum Illatha Palai*" (Desert that is for no one), discussed in the following section.

Palai in Contemporary Tamil Poetry

A Singaporean of Sri Lankan Tamil origins, Kanagalatha first migrated to Jaffna in 1983 when the Sri Lankan Civil War broke out and later settled in Singapore. In her poem, "Yaarukum Illatha Palai," Kanagalatha employs powerful imagery using a vision of Kali, the Sanskritic equivalent of the Tamil, *Kottravai*, who is the patron goddess of *palai*. It is noteworthy that *palai* is the only *thinai* with a female patron deity. She is considered the goddess of war, victory, fertility and hunters. The desert, which is so frequently in western cultural imagination correlated with "barrenness," is suffused with the generative power of the feminine.

The iconography of the goddess Kali has lent itself to feminist reclamations and alternative constructions of femininity. In Kali, the feminine is "powerfully terrifying."[12] While devotees see a mother figure, the deity, engaged in a battle dance against evil, is not domestic. Siting *palai* as the combative space between good and evil, Kanagalatha's poem is deeply metaphorical in its iteration. After defeating the *asuras* (demons) in war, a victorious Kali stands tall in *palai,* firmly renouncing war: "In the endless terrain / burnished by heat / buffeted by scorching sand storms." That the setting is *palai* can be gleaned from the description of the land and the references to the burning heat and sand storms making it clear that this is the region ruled by *Kottravai* or Kali, who, as the patron goddess, has sole dominion over it. Since in Hindu mythology this is where Kali has destroyed the male *asura* or demon, by extension, the desert quite literally becomes "no man's land"—a space imbued with feminine energy, power and control. However, notably, now that she has freed the land of malefactors, she is done with war. She sees no further need for violence. For her, war was a painful mission undertaken to rid evil and now it is successfully accomplished. Therefore,

She shakes each of her hands free.
Her weapons fall, registering sound.
Attuned to the music of warring weapons
the assortment of singing women
tie flags to
Kali's barren arms.
Her eighteen arms bear several songs.

The act of shaking her arms free of weapons is a particularly momentous one. In doing so she creates a kinship with those women, who, all their lives, have been forced to tolerate war. While war-making and 'war' per se are largely masculine pursuits, women die, face persecution, or live and are impacted by the ghastly consequences of war. Hence, the end and renunciation of war becomes a celebratory moment for women who rejoice with her.

As she sheds "shades of red," she "anoints herself in vibrant hues," tacitly signaling her wish to pursue a peaceable existence henceforth. This might appear incongruous since Kali is often considered the most aggressive of the goddesses in the Hindu pantheon. But her measured actions are validated by the women of *palai*, who stand in solidarity with Kali. Their war-scarred minds recognize that Kali alone has the strength to impede and negate patriarchy's ubiquitous, blithe and divisive narratives of nationalism and nationhood echoed in Benedict Anderson's depiction of the nation as an imagined political community where fraternity makes it possible for so many millions of people, "not so much to kill, as willingly to die for such limited imaginings."[13] Hence, it is not surprising that wherever Kali goes, what she is asked first is, "what is your country?" Consequently, there is deep deliberation in her action when,

In great pain,
she bound tightly
to the now tender arms
national flags.

Here, flags that are symbols of nationhood function as active symbols of war. In her final act of renunciation, Kali's hope for peace overrides patriarchal institutionalization of violence:

With guns stacked
in her crevices
she abandoned them
in the deserted sea
bereft of habitat

Discarding the guns and national flags, Kali effaces the dominant discourses of war. The restituted land, "*yaarukkum illaatha palai*," can be translated as "desert, that is for no one." However, semantically, it can be cognised as "no *man's* land." Remarkably, despite the superficially desolate connotations it invokes, the poem's mindscape in *palai* is reimagined. With ideological connotations superseding geographic affiliations, it emerges as a place of greater hope and presents the possibility of a world that is free of war and its horrors.

It is important to acknowledge that while there are crucial differences that distinguish desert terrains from each other, what markedly sets them apart as a conceptual topography is the ability to embody singularity and a multiplicity at once. Literary deserts and deserts in creative imagination often articulate the anxieties of the time. They are also frequently reimagined as spaces of resistance, spiritual revival and liberation as evidenced in "*Yaarukum Illatha Palai*." The optimism evident in Kanagalatha's poem is in stark contrast to the sentiments expressed by other contemporary Tamil poets in the anthology edited by Chelva Kanaganayakam, *In Our Translated World* (2013). In "Drought" by Indian Tamil poet Raja Sundararajan, the cruel aridity of *palai* is perceived as the inevitable result of human actions that brings about gradual desertification. The tone is ominous in setting the scene of *palai:*

> The sky was merciless.
> The sun beat down on the earth
> spitting fire.[14]

The poet bemoans the fact that with the wells running dry, the roaring waterfall now a barren rock, and the still air void of the benediction of breeze "the house stands wrapped in grief." This poem bespeaks the sorrow of a civilization. The gradual desertification[15] of the Tamil land soaks the landscape and its residents in sorrow, in place of water. This warns of a future where the five *thinai* landscapes could well become reduced to a single, arid landscape, *palai*. As the global climate crisis escalates rapidly, having particularly alarming ramifications for the Global South, literary, philosophical and cultural critiques of deserts as well as various kinds of barren landscapes can help promote a more holistic understanding of the environmental crisis at a time when human activity has a significant impact on the planet's climate and ecosystems. As an imaginative space, the portrayal of the desert in this poem also finds resonance with the themes of war and pillage. We see that the desert serves as a powerful trope in modern Tamil poems on the Sri Lankan Civil War.

A well-known Sri Lankan Tamil poet and political analyst, S. Sivasegaram's poem "War and Resolution" clearly aligns with *puram* poetry and grapples with a dark paradox: "if peace solves nothing, why shouldn't war be welcome?" The poem begins on a promising note of contemplation and peaceful compromise, where the emphasis is on negotiation. But this promising sentiment soon descends into doubt: "For a moment/this felt right;/the next moment/it felt wrong."[16] For they realize that peace too does not offer a solution. Finally, the casual remark of a gun-toting boy that all things are "resolved by war" sounds more equivocal than inflammatory to the speaker. The stark war scene is reminiscent of the desolate *palai thinai* where hope is waning, and to the forlorn speaker everything is appears futile and meaningless. This aridity is reinforced by the poetic voice in "*Sithanthan*" (philosopher), who declares that he has been reappearing since time immemorial only to repeatedly register the changing landscape from lush vegetation to "burnt out forests" and "ash coloured mounds," both bespeaking gradual desertification. This infertility of the land occurs in tandem with a barrenness of the soul. The extreme anguish caused by the distended civil war of Sri Lanka is captured in another popular Sri Lankan Tamil poet, S. Vilvaratnam's "The Alienated Land":

> With the coming of the Outsiders…
> rooted lives were plucked,
> all the people left…
> the village looked
> like a land
> emptied of the floodwaters
> of life.[17]

Just as in *palai thinai* the forlorn lover feels bereft of love, here, the "outsiders," the alien looters, have depleted the village of its life-giving energy, both material and emotional. Thus, the prominent motif of *palai thinai* remains centered on pain and loss.

The idea of barrenness, both metaphorical and literal, casts its dark, persistent shadow over *palai* in contemporary Tamil poetry, resonating to the tumult of civil war, refugeeism, and emotional deprivation which are the immediate concerns of (especially, Sri Lankan) Tamils. While Kanagalatha's poem puts a positive spin on the bleak, war-torn landscape, the poems from *In Our Translated World* remain enveloped in pessimistic ruminations on war and loss. This polyphony in the conceptualizations of the desert is significant. Both as a hostile wilderness that encompasses eschatological concerns such as war, death and destruction, and as a space embodying feminine power and liberation from patriarchal grand narratives glorifying war, the desert signifies both a dystopian past and present as well as a veritable "no man's land" where war is denounced.

Palai in Prose Fiction

In contemporary prose fiction too, *thinai* has preoccupied writers, who have used it in innovative ways. In her short story, "*Thinaigal*" (*Thinais*), Kamala Devi Aravindan uses *thinai* to register her angst against patriarchal-capitalism. A prolific Singaporean author, Aravindan has published five novels, several plays, and short stories. Conjoining landscape and mindscape and describing a liminal space that hangs suspended between reality and dream, the story highlights *thinai*'s capacity to accommodate fluid identities. While the story is set in real geographical locations in Singapore, such as Bukit Timah Forest and Bishan, the landscape descriptions, replete with the five geographical *thinai*[s], appear to be more part of a dreamscape. Characters like Mullai and Marutham have been named to correspond to *thinai* regional labels (*mullai* and *marutham*, respectively), making the story highly referential.

Value dualisms are at the heart of the narrative as it foregrounds a primary human-Self/nature-Other binary between the protagonist, Rajasekar, and the natural world. Rajasekar embarks on a brisk stroll through the Bukit Timah forest one day and encounters a strange adventure. He conforms to an anthropocentric code that assures him of a transcendent position above nature, which he perceives as existing solely for his benefit. The narrator clarifies that he is an alien in these natural surroundings since he has ventured into the forest more or less under duress in order to keep himself fit for a proposed trip to the Himalayas, where his carbon-heavy modes of travel might occasionally be interspersed with short walks up little hillocks. Finding himself suddenly in a forest of fruit trees, not surprisingly, he proceeds to pick one. He is startled by a female voice that reprimands his action. The voice, we learn, belongs to Mullai, who appears to be the *genius loci* of the place, advocating an ecocentric ethos. She discloses that she has only begun residing recently in the forest. "Where did you allow me to live in the city? I was previously living in Shenton Way. But in the name of urban redevelopment and deforestation, you drove me here."[18] Her words reveal that she is both imbricated in and also bereft of human predatoriness. Mullai functions as symbolic of (an aspect of) nature. Her accusing Rajasekar of deforestation and urbanization draws the Self/Other binary along the lines of "exploiter" versus "preserver" of earth's resources. As they squabble over whether or not he has the right to the fruit, their contrasting perspectives get firmly established. While Rajasekar embodies an unrelenting instrumentalism, where natural objects are viewed as existing solely for his use, Mullai staunchly upholds the agency and autonomy of nature. This tussle between human consumerism and the rights of an agentic nature becomes a prominent theme in the story.

As Rajasekar traverses the five *thinai*[s], his own exploitative nature is repeatedly contrasted to nature's generosity. However, he continues his truculent behavior. Finally, he reaches the dry expanse of *palai* land. As if to signal the chaos of modernity, he finds Mullai and her lover Marutham, being physically intimate in this arid land. To the readers, the lovers' affinity is a natural extension of their alignment with nature and the respective landscapes. Marutham is an ardent champion of nature. However, Rajasekar, more preoccupied by cultural codification, is bemused by this contravention of *thinai* parameters and exclaims, "surely illicit love is categorized under *Kurinchi*. Why are they practicing it in *palai?*"[19] This comment brings to the fore a tension that had been lurking beneath the surface narrative, namely modernity's rupture with the past. As the modern urbanite, Rajasekar, the city-dweller, is sharply cast against the upholders of traditional/ancient values and spaces, Mullai and Marutham, the narrative illuminates the gradual desanctification of Earth, a theme that Carolyn Merchant explores in *The Death of Nature*.[20]

Western philosophers such as Neitzsche, Heidegger, Levinas, Deleuze, Guattari and Baudrillard repeatedly deploy the image of the desert in their critique of modernity (Tynan 7).[21] In their philosophical disquisitions they use wasteland imagery as consonant with the dysfunctionality of modernity. Similarly, in "*Thinaikal*," the ideological disconnect between Rajasekar on the one hand, and Mullai and Marutham on the other, stands for that idea of the 'modern' as exhibiting a decisive rupture with preceding, sanctioned frames of references. The resultant tension makes *palai* the land where the final confrontation between the ancient and the modern is enacted. Bruno Latour suggests that with the word 'modern' "we are defining, by contrast, an archaic and stable past…it designates a break in the regular passage of time, and it designates a combat in which there are victors and vanquished."[22]

This fits the premise of the narrative where Rajasekar's egocentric arrogance stands in stark contrast to the traditional values that Marutham and Mullai uphold, such as respect for nature and cognizance of human embedment in nature. In his insistence that generative nature has no place in the desert land of *palai*, he reiterates his solipsism and self-centeredness that delude him into overestimating his power over the natural world. Therefore, it is apposite that at that instant in *palai* land, while he stands mocking Mullai and Maruthan, he should feel the pulse of a sudden, overwhelming power envelop him. Shaken, he swerves around to be confronted by a gigantic form with many arms and dreadlocks, standing in his path. This is an iconographic allusion to *Kottravai*, the patron deity of *palai.* Alienated from the power of nature, trained to view nature as mere resource, after the initial moment of panic, Rajasekar apprehends this great natural power as only a big tree standing in his path. Here, it appears that nature, landscape, and the feminine spirit are banded together against the predatory male. Mullai, standing by, reprimands his lack of reverence for the 'elder' tree. Rajasekar scorns such an obsolete view of nature. But, ironically, in the final analysis, the man does need Mullai's help, after all. The lost Rajasekar begs Mullai to show him the way home. She agrees on the condition that he gives her a present.

Typical of his chauvinistic attitude, he assumes that she is alluding to sexual favors. He blithely agrees. Finally, after she leads him back through a tortuous route and demands her prize, he embraces her roughly and plants a kiss on the irate Mullai. The story ends with her resounding slap ringing in his ears for days to come as he is once again "safely" ensconced in his house in the city, reliving that strangely passionate embrace: the symbolic embrace of the rapacious male and "feminized" nature, which ends in a stinging reprimand.

The short story draws attention to the tensions and conflicts that arise when the modern is juxtaposed with the traditional. Here, modernity is perceived as alienation from nature; it implies a departure from the past and is seen as embracing a utilitarian and exploitative attitude towards nature. The text signals that in the current era, *palai* is not merely an arid land of pain and loss, but nature—communicating through the representational forms of Mullai and Maruthan—has learnt to exist and to explore emotions, including love in this parched land. More importantly, *palai* is also reimagined as the land that teaches the human male to 'fear' nature again, even if this is only a momentary lapse. *Palai*, the scene of Rajasekar's intimidation, is revealed as the last chance given to the androcentric/anthropocentric man to recalibrate his actions and attitudes. It seems, however, that this lesson has been wasted one.

Palai is also presented as a landscape that prompts reflection in the novel *Nilam Poothu Malarntha Naal* (2015)[23] (*The Day the Land Bloomed and Blossomed*). The highly acclaimed novel by the Malayalam writer, Manoj Kuroor was translated into Tamil by the award-winning K.V. Jeyasri. It unfolds from triple narrative perspectives of a father, daughter, and son, who belong to the clan of the *panar* (nomadic bards) the ancient Dravidian equivalent of the medieval wandering troubadours of Europe. The extensive narrative is set in the earlier part of the *Sangam* era. The text offers richly researched references to the famous Chera, Chola and Pandiya kings of that period and to the *Sangam* poets and bards, such as Paranar, Kapilar and Avvaiyar, whose poetical works are still admired and read by Tamils. Finally, the novel unfolds through the five *thinai* lands, intimately describing these in lyrical prose that mimics the cadence of ancient Tamil poetry.

Since the *panar* lead an itinerant life, the description of the five *thinai* lands unfolds realistically as they move from one kingdom to another. As though to induct us into the vicissitudes of life in the arid *palai*, the preceding narrative ends with the death of the first narrator, the father. This *panar* is killed inadvertently by the actions of his own son, Mayilan, in search of whom he had set out to various kingdoms with his family. The dark mood of untold and unresolved problems of the heart that often forms the staple of *palai thinai* is patent here. The daughter, Sithirai, the ensuing narrator, picks up the tale. Mourning the death of her father, Sithirai is courted by an enigmatic soldier, Makeeran. He wooes her and in the best tradition of *kalavu,* illicit love (widely prevalent and sanctioned in ancient Dravidian literature) Sithirai leaves her clan of bards to follow Makeeran through *palai* land. True to the sentiment of pain and loss linked to *palai,* Sithirai's lover deserts her, and she is left wondering if he will indeed return for her. Eventually, Makeeran returns but his distant, enigmatic attitude towards her is a source of constant worry for Sithirai. Seated behind her inscrutable lover on his horse, Sithirai rides through *palai* land. The classic description of *palai thinai* with its thorny shrubs, extreme heat and dry terrain, is further embellished by the presence of bandits, who roam the land in search of gullible prey. When a group of them converge on Sithirai and Makeeran, she is terrified. But her lover appears calm and, surprisingly, seems on quite familiar terms with the bandits. They reach the next kingdom, which he claims is his home and she discovers that he is the captain of the king's army. But increasingly, he leaves her alone at home and appears to have no time for her, foregrounding the Palai theme of separation.

Sithirai soon finds out that Makeeran is not only the murderer of her missing brother, Mayilan, but he is also living with another woman in an adjacent kingdom. Her family entreats her to return with them to rejoin their *panar* clan, which is planning on returning home. However, the shocked and disillusioned Sithirai decides that it is time she stood on her own two feet. She expresses her resolve to follow in the footsteps of Avvaiyar, the female bard from the Sangam era, who roamed the Tamil land unfettered and unafraid because her learning earns her immense respect and opens even royal doors. Thus, in this contemporary novel, the traditional theme of *palai* land, which features pining, bereft women, hopelessly longing for their lovers to return, is replaced by a more robust and feminist theme, where the scorned and deceived female protagonist embraces self-sufficiency, resilience, and resourcefulness. She is no longer dependent on a male for her happiness or subsistence establishing that education and autonomy appear to be the way forward for women.

In the third narrative, told by Mayilan, Sithirai's older brother and the lost son whom the *panar* clan had set out in search of at the outset, Mayilan recounts his adventures: how, running away from his poverty-stricken home as a young boy, he is captured by a group of bandits and becomes part of the *maravar* or the clan of brigands, who inhabit the arid *palai* land. Here, he becomes adept at the art of robbery, looting and murder. Ultimately, he is captured by a bard and led away from *palai* to a fertile kingdom. There, the ambitious Mayilan becomes a foot soldier and climbs the ranks to become the king's personal guard, and he becomes a part of the soldier clan.

After several (mis)adventures, trials and tribulations, Mayilan is befriended by Makeeran, who is a spy. Mayilan is persuaded to become one too. He accidentally slays his own father who has come with his *panar* clan to sing at the king's court. Mayilan then retreats in anguish and Makeeran takes him to another bandit clan. Seeing he is exhausted and agonized, the clan care for him as one of their own. Residing in *palai,* but without participating in looting and murder, Mayilan lets the void landscape and its solitariness soothe his troubled mind. He resolves to find a more peaceful way of life. Once again, *palai* is shown to offer respite and even to encourage reflection. The desolation of the land, it is implied, nurtures introspection.

Thus, in *Nilam Poothu Malarntha Naal*, though the arid land of *palai* is presented as a bleak, inhospitable terrain, it is also a place where even the bandits can demonstrate a rugged solicitude and adhere to their own code of honor. It is shown as a land of rough and ready justice. It is also a land that teaches both men and women, endurance, *sitzfleisch*, resolve and indeed, discernment.

In the novel though Sithirai is centered, the action remains a male prerogative. In the visual texts we examine below, however, the state of subalternity gets reinscribed into *palai* in entirely different ways as women become the central agents in the diegesis and fight to bring about change.

Palai in Film

Thaneer Thaneer[24] (Water…Water) (1981) is a Tamil film, directed by the well-known director, K. Balachander. It is based on a 1979 play written by Komal Swaminathan. Critically acclaimed, the film won numerous awards. The story is set in a drought-ridden village called Athipatti, which is facing acute water scarcity. As the villagers navigate the political minefields of corruption and government apathy, through the collective action depicted in the film what is emphasized is subaltern and female agency. The female protagonist spearheads a movement that transcends the boundaries of caste and gender while the arid *palai* land is imbued with material agency and acts as a relational force. The aridity of the land is shown to impact not only lives and livelihoods but the philosophical imagination of the collective to create new awareness.

The film has three main protagonists, one female and two males, and opens with the female protagonist, Sevanthi making the long and laborious ten-mile walk through the dry, arid lands of her village and its neighboring areas to fetch water. She carries her baby in a cloth sling. Her journey back with the additional load of multiple water pots is even more arduous. The shots pan between the hot sun beating down on her, the parched, dry vegetation and cracked earth, intensifying the terrible physical trauma she endures to make the journey back and forth. Her walk back to the village also reveals the extent of the villagers' desperation for water. On her way, she sees two laborers, crouched on all fours, drinking water from a dirty puddle. Perturbed by this, Sevanthi offers them water from one of her pots. Further down, she is coaxed and cajoled by the temple priest into offering a pot of water to appease the village deity. The priest promptly sneaks away the pot filled with water to his wife. Exercising many easily identifiable instances of agency right at the outset, the significance of Sevanthi's character is established. She is the first in the village to proactively venture out of her village to get water. In this, she displays self-reliance. A lone crusader, she contends with the derision of others while the rest of the villagers are passive onlookers and sufferers.

It is noteworthy that even through the visual medium, the diegesis remains faithful to the primary tenets of the *thinai* poetic code. The traditional association of the arid landscape of *palai* with separation plays out in *Thaneer Thaneer* as well. Sevanthi's husband, Alagiri, has migrated to the city in search of work. The film touches upon the issues of urban migration by highlighting the problem of lack of livelihood opportunities associated with arid regions. The description of the village's desiccated landscape contributes to the rich symbolism and satire employed in the film, focusing on the disenchantment of the working classes with the politics of developmental discourse and government inaction. For instance, on their way back from collecting water, Sevanthi and Vellaiswamy, a newcomer in the village, talk about a leafless tree that has the flags of different political parties tied to its branches. Vellaiswamy remarks, "So many flags for such a small village? Can the tree hold them all?" Sevanthi answers with a wry laugh, "That's why the tree is dead." In the last scene of the film, we see the flags transferred to a paddy field, covering it up entirely. The flags are symbols of political power and its control over natural resources and their distribution. These symbolic references, woven masterfully into the screenplay, expose the utter failure of bureaucracy and the government machinery in addressing the needs and aspirations of the subalterns and the working classes. The film goes on to depict how, through cooperative methods, the village gets together to take matters into their own hands. However, the two male protagonists, Vellaiswamy and Kovalu, mobilize political action in markedly different ways. Vellaiswamy, a convict on the run, is wanted for the murder of an oppressive local landlord. With the help of Vathiyar, the village school master, he begins a co-operative scheme to transport water in a cart from a pool ten miles away and later rallies the villagers to construct a canal to solve the water crisis. Kovalu, on the other hand, adopts a more radical method to confront the apathy and greed of those in power. His communist leanings are made apparent by the red neck-cloth that he sports throughout the film. He advocates open agitation and towards the end of the film takes recourse to arms.

In a talk organized by The Institute for South Asian Studies and the Tamil Studies Initiative at UC Berkeley, S. Shankar presented an insightful reading of the paradigmatic exchange between the "regimes of development and the agency of the poor." Shankar points out that the two male protagonists represent two different types of agency in questioning state-led development. Vellaiswamy, as Shankar posits, represents a "cessation from the regime of development through reliance on a sustainable economy of subsistence." Shankar identifies Kovalu's interventions as "confrontational agency"—one that uses physical force to coerce the government to distribute resources to the needy. Thus, the two male protagonists present self-reliance and open dissent as two means to an end.

Several attempts to get the administration to intervene are of no avail. The petition the villagers submit passes the hands of several government officials and reaches a dead end, exposing the ineffectiveness of institutionalized poverty-alleviation programs. Vellaiswamy exhorts the villagers to take control of the situation by encouraging them to build a canal themselves to bring water to the arid lands. Moved by the entreaties of the protagonists and fueled by their own desperation, the canal work undertaken by the villagers progresses steadily. Here, the barren land becomes a unifying force. It enables the villagers to put aside differences of caste and creed to harness the power of the proletariat. The film makes a compelling statement against vote bank caste politics. When unscrupulous politicians leverage caste to ask for votes, the villagers discuss who to vote for—Nayakar or Reddiyar? They realize that neither candidates from these 'upper-castes' have cared for their wellbeing or done anything to help with the water crisis. The village decides to boycott the elections, making a powerful political statement.

Sadly, their hard work with the construction of the canal does not reach fruition. An officer from the Public Works Department stops the villagers from finishing the work, on the grounds that they are engaging in unlawful acts on unassessed land which is the property of the government. Violence ensues and the villagers are shot at by the police. In the film's tragic ending, Vellaiswamy dies of thirst and Kovalu decides to become a part of an armed insurgency. While the villagers abandon the barren land, Sevanthi watches the sky relentlessly in hope of rain. The ending, while dystopian and stopping short of any redressal for the plight of the villagers, makes some powerful political statements on the discourse of development in postcolonial India. While it captures how the land drove the people into sheer desperation and plight, it also highlights its power to unify. Here, the drought-ridden land and the water crisis enable the villagers to explore their agentive potential. The landscape character of *palai* enables the creation of a biopolitical field in which the give and take of power is located. *Palai* becomes the site and the enabler of collective civic action—the strength of which is at the heart of *Thaneer Thaneer*'s didactic message.

The narrative begins with the display of the female protagonist's unassailable spirit in fighting the impact of the barren land and ends with her hopeful longing for a future in which she is liberated from the fate that her *palai* domicile has subjected her too. While she fights to alter the village's landscape through the canal, the land and those that have power over it seem to win the battle. Yet, in an act of powerful defiance, she neither abandons the land that has failed her, nor does she give up hope of making it more habitable.

In a similar vein, in the film *Pennadi* (2011) by Indian poet, filmmaker and actor, Leena Manimekalai, we see the female lead direct the path of the diegesis to present an empowering testimony of feminist power. *Pennadi* focuses a critical lens on the construction of female sexuality in Tamil poetry. As a leftist, Dalit,[25] lesbian, activist, Manimekalai has braved the ire of the conservative right wing, having received death threats and suffered book burnings. A prolific filmmaker, she has directed and produced ten documentaries and a feature film. She has published poetry on a range of topics, including sexuality, caste, and the plight of the dispossessed.

Deeply feminist in its persuasions, the title "*Pennadi*" is a thought-provoking one, encapsulating the polysemy and playfulness of the text that belies its serious intent to challenge entrenched and essentialized gender constructions. "*Penn*" in Tamil means woman and "*naadi*" means to seek or search. *Naadi* also means pulse. So the title could mean "in search of a woman," or "the essence of a woman." Alternatively, it also hints at another coinage where *penn* and *kannaadi*, meaning mirror in Tamil, have been combined to coin a new word, which could mean "the mirrored woman" or "the self-reflexive woman." However, the film's English title is "My Mirror is the Door." All these meanings and connotations can coexist and capture the layered texture of the screenplay.

Through a dialectical interaction between ancient and contemporary women poets, the film recontextualizes and reframes *thinai,* foregrounding its versatility and its usefulness as a fluid identity conception in representing female sexuality. Drawing attention to women poets who were a very strong presence in *Sangam* poetry, Manimekalai draws upon the woman-power from the past and harnesses it in the personalized journey portrayed in the film. The introductory shots of the documentary contextualize the feminist significance of *Sangam* poetry by discussing the high levels of female literacy and creativity during this time. Women poets like Avvaiyar, Velliveethiyaar, Nappasalaiyaar, Varumulaiyaarithi, Nedumpalliyaathai, Kachippaedu Nannakaiyaar and Parimakalir are named and their poems also narrated intermittently throughout the diegesis. *Pennadi* seeks to promote ecofeminist awareness by disrupting problematic contemporary constructions of female sexuality. The film foregrounds the female protagonist while virtually excluding the male. The male lover, presented only as a consort following the female protagonist, appears only in passing and is absent in most *thinai* landscapes, while the woman moves laterally from one *thinai* to another, challenging and reversing conservative gender hegemonies. For the purposes of this chapter, we will focus on the female protagonist's traversal through *palai thinai.*

The screenplay flows on, faithfully presenting the landscape and corresponding conduct codified under *thinai*. Thus, in *kurinci*, the protagonist is shown as united with her lover; in *mullai*, she awaits in vain for her lover to arrive; in *marutham*, she is shown as quarreling with her lover, and in *neythal*, she is pining for her lover. Finally, in *palai*, she is shown as separated from him. Carefully selected verses from *Sangam* poets are displayed intermittently to elucidate the associations between land and conduct. For instance, Velliveethiyaar's verse 27 and Avvaiyar's verse 28, both from *Kurunthokai*, an anthology of *akam* poems, allude to lovers' separation in *palai* and the woman's intense physical yearning. Avvaiyar's poem captures the angst of the forlorn lover who is unsure of how to express her longing:

> Should I bang on the door?
> Should I tear it down?
> What should I do?
> Should I scream out loud?
> To announce to this village, while it sleeps unmindfully,
> My growing sexual desire, that rises rapidly with this intolerable breeze.[26]

Evidently, the use of uncensored, explicit, sexual vocabulary used by these ancient women poets of the second century BCE is the highlight of these poems. Manimekalai also uses Velliveethiyaar's verse (*Kurunthogai* 27 discussed above) that explicitly foregrounds female desire. These classical poems of female sexual yearnings are followed up by Manimekalai's own poem on masturbation.

> My vagina pleaded for some love,
> I stitched it a hat with purple silk,
> Combed the hair into a rare flower…
> [I] caressed my pearl like it was hailstone fruit,
> My ululating vagina caressed me with a rain shower.
> We danced, one to one, for ages.[27]

This suggests a moment of liberation for the yearning lover of *palai thinai*, as she achieves sexual fulfillment even in the absence of her lover, highlighting the empowerment and control in the act of self-pleasuring. The female protagonist's separation from her lover is no longer an impediment to sexual gratification. Thus, *palai* becomes a space of self-discovery, self-reliance and bodily autonomy, taking the power of sexual expression in *Sangam* poetry a step further by using *Thinai* to defy and subvert patriarchy and assert female power. Reinventing *palai* in this manner, Manimekalai reframes *thinai* as an ecofeminist tool of resistance to align with female emancipatory voices and challenge mainstream repressions of female sexuality.

Oscillating between *akam* (interior/love/body) and *puram* (war/land), the rest of the narrative takes the audience through the movement of the two lovers through various *thinai*[s], juxtaposed with evocative scenes of war and protest representing contemporary human rights agitations. Using Sri Lankan Tamil poet, Sivaramani's poetry, Manimekalai links *akam* and *puram* to show how women's bodies (through violation and rape) have been viewed and used as "landscapes" in war. In a momentous shot, viewers are presented four televised events simultaneously via four separate but adjacent television sets. The first shows the finale of a Miss Universe pageant; the second, a protest march against violence by Kashmiri women in purdahs; the third television showcases Irom Chanu Sharmila, a civil rights activist and poet from the Indian state of Manipur, hospitalized and weak from her hunger strike, protesting against the rape of Manipuri women by Indian soldiers and the last showing naked women holding slippers outside the army headquarters in Manipur protesting against their molestation by soldiers.

In *Pennadi*, Manimekalai's excavation of *akam* and *puram* poems by *Sangam* women poets links them to her own poetic pieces and reveals how [human] emotion, sexual and gender politics, and *thinai* have come to be entangled in Tamil culture through the millennia. She traces female sexuality through 2,000 years of Tamil literary history connecting it to nature and ecology. This is much in line with Ariel Salleh's observation that "[i]n a time of ecological crisis, ecofeminists worldwide have become agents of history/nature. They give voice to a subversive politics, aware of its own situatedness and transitionality."[28] *Pennadi* makes visible women's agency in exploring both sexuality and ecology through appropriating *thinai* for feminist purposes. Manimekalai challenges the influence of patriarchal creeds that grip ecological and sexual ideologies, as well as the gender enculturation of Tamils that was a fallout of colonialism.

Conclusion

In this chapter, using ecocritical and ecofeminist perspectives, we have examined the works of contemporary Tamil writers who use *palai thinai*—depictions of dry, drought-ridden wastelands—in their texts to reimagine a space in which oppressive, hegemonic structures and dominant discourses are negotiated and challenged. We foreground ways in which dystopian and apocalyptic visions of deserts and arid regions, perpetuated by Western traditions, are conceptualized in these texts where *palai* becomes a significant ecocritical concept that is profitably accessed to seek empowerment and create spaces of resistance in response to patriarchy, caste-based domination and other forms of institutionalized oppression. A diverse corpus of contemporary Tamil texts from multiple genres has been examined to show how these narratives engage in productive ways with the *Sangam* version of an ecological society where humans do not seek to transcend natural conditions and respect ecological boundaries.

Notes

1 Bending, Stephen, "Literature and Landscape in the Eighteenth Century," *Oxford Handbook Topics in Literature* (Oxford University Press, 2015), https://doi.org/10.1093/oxfordhb/9780199935338.013.133.
2 Martin Döring and Francesca Zunino, "NatureCultures in Old and New Worlds: Steps towards an Ecolinguistic Perspective on Framing a 'New' Continent." *Language Sciences 41*, Part A (January 2014): 34.
3 Tynan, Aidan. "Introduction," in *The Desert in Modern Literature and Philosophy: Wasteland Aesthetics*, 1–5 (Edinburgh University Press, 2020), 2. https://doi.org/10.1515/9781474443371-003
4 *Sangam*, in Tamil, means literary academy. Literary historians trace three Sangam periods but there is no surviving literary output from the first two. So, when reference is made to Sangam literature, it usually points to the third Sangam period. During this time, the Tamil kingdom was ruled by three major dynasties: Chera, Chola and Pandya. This period has been identified as one of great intellectual fervor. Significantly, there were substantial contributions from female poets.
5 *Tamilagam* refers to the territories of the present-day Indian states of Tamil Nadu, Kerala, parts of Andhra Pradesh, parts of Karnataka and northern Sri Lanka.
6 M. John Britto, "Sangam Landscapes and Thing Theory: A Study with Reference to Kurunthogai." *Forum for World Literature Studies* 9, no. 4 (Knowledge Hub Publishing Company Limited), 705.
7 The first.
8 Kernel or embryo.
9 Essence.
10 For more details on rules of *thinai*, see Ilakkuvanar (1963) and Takahashi (1995).
11 Unless otherwise stated, all translations are the authors' own.
12 Dalmiya, Vrinda. "Loving Paradoxes: A Feminist Reclamation of the Goddess Kali." *Hypatia* 15, no. 1 (2000): 125. http://www.jstor.org/stable/3810514.
13 Benedict Anderson, *Imagined Communities: Reflections on the Origin and Spread of Nationalism* (Verso, 1991), 7.
14 Raja Sundararajan, "Drought," in *In Our Translated World*, ed. Chelva Kanaganayakam (TSAR Publications, 2013), 117.
15 For more on the distinctions between deserts and desertification, see Chapter 9 in this volume by Veronika Kratz.
16 S. Sivasegaram, "War and Resolution," in *In Our Translated World*, ed. Chelva Kanaganayakam (TSAR Publications, 2013), 67.
17 S. Vilvaratnam, "The Alienated Land," in *In Our Translated World*, ed. Chelva Kanaganayakam (TSAR Publications, 2013), 79.
18 Kamaladevi Aravindan, "Thinaikal," *Tamil Murasu*, Sunday Edition. trans. Chitra Sankaran (Tamil Murasu, 22 February 2015), 12.
19 Aravindan, "Thinaikal," 12.
20 Carolyn Merchant, *The Death of Nature: Women, Ecology, and the Scientific Revolution* (HarperOne, 1990).

21 Tynan, Aidan. "Desert Desire," in *The Desert in Modern Literature and Philosophy: Wasteland Aesthetics* (Edinburgh University Press, 2020), 7. https://doi.org/10.1515/9781474443371-003
22 Bruno Latour, *We Have Never Been Modern*, trans. Catherine Porter (Harvard University Press, 1993), 10.
23 Manoj Kuroor, *Nilam Poothu Malarntha Naal,* trans. K. V. Jeyasri (Vamsi Books, 2015).
24 *Thaneer Thaneer*, directed by K. Balachander (Kalakendra Movies, 1981). https://youtu.be/L0Htv5mcwo0
25 Dalit, meaning broken or scattered, is an affirmative political term adopted by the so-called 'untouchable' castes for self-reference.
26 *Pennadi*, directed by Leena Manimekalai (2012; Public Service Broadcasting Trust India, 2015), https://www.youtube.com/watch?v=DwXPUlmAzgM.
27 *Pennadi*, directed by Leena Manimekalai (2012; Public Service Broadcasting Trust India, 2015), https://www.youtube.com/watch?v=DwXPUlmAzgM.
28 Ariel Salleh, *Ecofeminism as Politics: Nature, Marx and the Postmodern* (Zed Books and New York: St Martin's Press, 1997), ix.

Bibliography

Anderson, Benedict. *Imagined Communities: Reflections on the Origin and Spread of Nationalism*. Verso, 1991.

Aravindan, Kamaladevi. "Thinaikal." *Tamil Murasu Daily, Sunday Edition*. Translated by Chitra Sankaran. MediaCorp, February 22, 2015, 12.

Balachander, K. dir. *Thaneer Thaneer.* Kalakendra Movies, 1981. https://youtu.be/L0Htv5mcwo0

Bending, Stephen, "Literature and Landscape in the Eighteenth Century." *Oxford Handbook Topics in Literature*. Oxford University Press, 2015. https://doi.org/10.1093/oxfordhb/9780199935338.013.133.

Dalmiya, Vrinda. "Loving Paradoxes: A Feminist Reclamation of the Goddess Kali." *Hypatia* 15, no. 1 (2000): 125–150. http://www.jstor.org/stable/3810514.

Döring, Martin and Francesca Zunino. "NatureCultures in Old and New Worlds: Steps towards an Ecolinguistic Perspective on Framing a 'New' Continent." *Language Sciences* 41, Part A (January 2014): 34–40.

Ilakkuvanar, Singaravel. *Tolkappiyam (in English): With Critical Studies*. Kural Neri Publishing House, 1963.

Kanagalatha, K. "Yaarukkum Illaatha Paalai." Translated by Chitra Sankaran. *Tamil Murasu Daily*. MediaCorp, 2014.

Kanaganayakam, Chelva. *In Our Translated World.* TSAR Publications, 2013.

Latour, Bruno. *We Have Never Been Modern*. Translated by Catherine Porter. Harvard University Press, 1993.

Manimekalai, Leena. dir. *Pennadi*, 2012. New Delhi: Public Service Broadcasting Trust India, 2015. https://www.youtube.com/watch?v=DwXPUlmAzgM

Merchant, Carolyn. *The Death of Nature: Women, Ecology, and the Scientific Revolution*. HarperOne, 1990.

Murugan, V. *Tolkappiyam in English.* Institute of Asian studies, 2000.

Salleh, Ariel. *Ecofeminism as Politics: Nature, Marx and the Postmodern*. Zed Books and New York: St Martin's Press, 1997.

Shankar, S. "Poor Stories: Interrogating Discourses of Development in Tamil Film and Play 'Thaneer, Thaneer'" (webinar, The Institute for South Asia Studies and the Tamil Studies Initiative at UC Berkeley, Berkeley, 3 November, 2020). https://fb.watch/bTY0fJ49gl/

Sivasegaram, S. "War and Resolution." In *In Our Translated World*, edited by Chelva Kanaganayakam, 67. TSAR Publications, 2013.
Sundararajan, Raja. "Drought." In *In Our Translated World*, edited by Chelva Kanaganayakam, 117. TSAR Publications, 2013.
Takahashi, Takanobu. *Tamil Love Poetry and Poetics*. E. J. Brill, 1995.
Tynan, Aidan. "Introduction." In *The Desert in Modern Literature and Philosophy: Wasteland Aesthetics*, 1–5. Edinburgh University Press, 2020. https://doi.org/10.1515/9781474443371-003.
Vilvaratnam, S. "The Alienated Land." In *In Our Translated World*, edited by Chelva Kanaganayakam, 79. TSAR Publications, 2013.

6

YAWP FROM ATOP THE DESERT SAND

The Poetics of Existentialism in *Jāhiliya* Poetry and Beyond

Alireza Korangy

> Thus, we see the common topics of their poetry to be domestic life, wars, heroic deeds, martial triumphs, travels, camels, horses, weapons, chase, love, reminiscences of old associations, hospitality, glory and genealogy of the tribe, panegyrics of noble personages and chiefs, records of their patriotic and virtuous deeds done for the good of their tribes, acknowledgment of their obligations, elegies, embodying posthumous recollections and commemorations of the virtues of deserving merits in pro-portion to their deserts. Precepts of sociology, political views, philosophical doctrines, maxims and proverbs were not lacking; but they were mere results of a direct observation of the objects of nature and of a deep contemplation of humanity in its simplest aspect.
>
> Clement Huart, *A History of Arabic Literature* (1903)

The Bucolic nature of Arab existence in pre-Islamic Arabia has been subject to a plethora of exegesis.[1] The term *Jāhiliya* (age of ignorance), as first impressions go, brings to mind a wild existence devoid of thought and profundity. This term was originally coined to point to a lack of access by these pre-Islamic era folk to the 'divine truth', i.e., Islam and the prophet's teachings, however, it erroneously and eventually came to signify ignorance at large. This line of thinking about these nomadic, polytheistic peoples is certainly commensurate with the image of the desert as a totality devoid of civility—and hence intellectuality—and moreover implies an absence of imagery, per se, at least to the extent that imagery is conceived as an essence with palpable, even sophisticated, dimensions and the sub-schemas that comprise it. The image, however, presents a stage for the will of humankind

DOI: 10.4324/9781003397212-9

to explore spirituality in other ways and modes exactly because of its otherwise 'horror vacui':[2]

> The will is innately capable of apprehending its own voiding in a generalised loss of meanings and aims. The ascetic ideal—the attachment to a 'spiritual' domain of values rather than to the physical here and now—emerges as a solution to this horror. Faced with the spectre of a nothingness of will, the will instead decides to will the nothingness of a spiritual reality.[3]

Part of what is considered 'ignorant' about the voice of this period (pre-610 CE)—which is the time period in the peninsula before the coming of Islam—is part of a post-rationalized notion that considered Islam as an awakening to a transcendent purpose—and as such construing those not privy to its teachings ignorant—and indeed lost. This, not then, and not now, cannot imply ignorance in an innate sense of the word. The poetry of the said folk 'of the kindred spirit' as Nicholson[4] eternalizes it, takes abreast truthful poetics symbolizing the environment in which they lived. These poets and poetesses of 'the barren land', from whom, incontrovertibly stem the reservoir of all poetic genius in the post-Islamic Middle East and Near East, were soundboards to an otherwise silence. Theirs, as an oral tradition, certainly lacked the complexities of later years of poetics we witness in post-Islamic traditions in the Persian-speaking world and the greater Arab-speaking regions in terms of aesthetics. When judged in terms of maturity, it was, on the hindsight, not refrained by conventions that often damage the intended conceit of a poet. It is undeniable that their poetry and poetics were the fledgling for the creation of a complex canon of rhetoric after the advent of Islam.

As the cities in the Arabian Peninsula were places where trade was commonplace, and were centers for public displays of devotion to one's divinity, the desert became the unique reflective stage for many.[5] The desert's void was an apt stage, a tabula rasa of sorts, for humanity's ever-present existential discourse and a nurturing cradle to ponder other philosophies and hermeneutics that behoove love, death, silence, and emptiness as profound junctures whereby human existence—and survivalist existentialism—becomes acutely highlighted; henceforth, prompting thought, rigorous vitriolic imagery, and a peculiar, yet profound, intricate Bedouin philosophy, which is considered unique due to the truth embraced by its poets. As the thematic dimensions of their genius went far and wide beyond the borders of Arabia, they served as platforms for many other poetic forms, i.e., the ghazal, but most importantly in many instances catapulted imaginations of the folk by making necessary the deep gaze, the layered image, and re-estimations of the otherwise normal:

> to Bedouin, as to Don Juan in Mexico, the desert is full of immanent 'power spots,' the landscape is mythologized, and is neither lifeless nor terrifying. But to the seminomad who must live next to it yet could not flourish on it, the

> desert's formlessness could suggest … the 'Wholly Other.' Ultimately, it is the discontinuity of the desert with the usual forms of life that could give the paradox of a concrete image of transcendence.[6]

The whole of what is known to us today as pre-Islamic poetry is otherwise known as the *Mu'allaqāt* (or hanging, or 'suspended'), a namesake which rates further discussions.[7] What follows signifies the role of the desert in understanding the human condition as pertains to the nomadic life. The following analyzes some of these expressions and as such (1) shows that absence is plentitude, both philosophically and poetically, and (2) highlights an oral tradition that stands as a testament to the Aristotelian notion of poetics as pertains to his concern with milieu as poetics.

The early platonic 'cave', under the rubric of the desert discursive, and the desert image, is plenty apparent in the poetry and poetics of these folk. To these poets, the desert was anything but a shapeless ad infinitum. It was a palpable personification of destiny which possessed, in the image of human, characteristics and functions, most importantly representing that vast unknown embodied in the vicissitudes of the Celestial Sphere (*falak*) for which the desert is thematically jettisoned in those later influenced by these early poetics:

> Night and day, I see you have emulated the way of the heavens:
> Every curtain you sow, you go on to tear down again.
> From my eyes, ruby red wine flows down cup after cup,
> Every time I would think you are remotely concerned for a miserly one such as me. You have the temperance of the celestial sphere, Therefore, just like the celestial sphere,
> Once in a while you swallow a hundred and release one in exchange.[8]

So, the platonic notion of the case, as aforesaid, is materialized in a poetic image whereby what one sees is not what is. At no point in the history of poetics can one create a divide between philosophy and poetry. In fact, poetry becomes the final resort for philosophy where coded language is always innately implied, and interpretations are thereby required. One cannot expect this of normal day-to-day language and as such many an intent by the speaker goes unnoticed. To that end, the poet here opts to use personification, whereby heavens take on human characteristics, i.e., 'every curtain you sow', 'you have the temperance of the celestial sphere', etc. As such, the desert becomes a stage, a papyrus if you will, where a story is told. Its infinite blank implies infinite tales of wonder.

What is most important is that the sound of the desert often encapsulates, figuratively, the merit of what happened there, i.e., what the desert determined, its functionality or rather its indispensability as a means to livelihood, and those who braved it. It did not promise prosperity. It was an obstacle course representing life, and a didactic theater demanding respect and awe: Labid, in a *qasida* describing the rural life and admired characteristics of his ancestors says, "In the rough grounds in the valley of Thahiboot up which he goes; *and although* its places of observation are empty, the fear is in the milestones".[9]

All in all, the desert served to inspire an active meditative mode of expression sounding human emotion for the *Jāhiliyya* poet. Although its poetic utilization becomes somewhat passive in the post-Islamic world and the later development—and usage—of that image in its poetry, its cultural significance remained embedded in the poetics of the region as a soundboard of reflection, although commensurate with the ecological and social dynamic particularities of each milieu.

In engaging the poetics of desert in Arabia, one is necessarily engaging the poetics of tribal life and tribal machismo. These poetics behoove gazes into their densely coded comraderies and its components, cross-tribal interactions, stern observance of revenge, the pastoral life, human-animal connections, romance, etc. Every macro aspect is indeed a part of a whole which embraces the concept of the desert; as it was ultimately the desert, and of course its attributes that set apace all that comprised the hermeneutic and ethos of the tribal persona, and all of its components.

Even the very early meters created by the *Jāhiliyya* were set apace in accordance with the movements of the camel; hence, poetry was not merely an expression, it was an existentialist journey deeply entwined with the concept of the desert:

> The long caravan-marches across the monotonous deserts, when the camel's steady swing bends the rider's body almost double, turning the unaccustomed traveler sick and diddly, soon taught the Arab to sing rhymes. He even noted, very soon, that as he hurried the pace of his recitation the long string of camels would raise their heads and step out with quickened pace. This creature, stupid and vindictive though it be, is sensitive, to some extent, to music, or, at all events, to rhythm.[10]

As such the rhythm, dictated by the co-hort, the camel, and supplemented by the ditty of the rider becomes the story of the journey, as though the cursive movement of pen on paper: pen (rider), the movement of the hand (the camel), the paper (the desert), and the poem (the journey).

Although most of these rhythmic expressions were in the form of *qit'a*s,[11] and *qasida*s, the themes adopted for each of these content forms far exceeded what was later to become appropriated in their poetical vertical axis: certainly, this was inadvertently done. The ethos, per se, of these Arab visionaries was best capitulated in the first 15–20 lines of their *qasida*s where their hyperintellections truly shone. These lines were early saplings of what was to become the compounding and ever-so-profound ghazals, a complicated genre deeply immersed in existential, antinomian, and woeful reflections.

Although the origins of the ghazal are still considered obscure at best, most place the earliest ghazal in the introductory amatory exordia (*tashbīb* and *nasīb*) of pre-Islamic poetry (*Jāhiliyya*) far pre-dating the more formal understanding of the ghazal. Embedded in the early Arabic ode, the *nasib/tashbib* often spoke to love lost, the vagaries of time and the transitory nature of life, often set upon the theater of the desert and movement thereabouts with the persona of the beloved at its core: a beloved whose affections were no longer within reach of the grief-stricken lover. The pre-Islamic poet Imru al-Qays's (d. 544)[12] lines on love and longing are famous:

> Halt, friends both! Let us weep, recalling a love and a lodging by the rim
> Of the twisted sands between Ed-Dakhook and Haumal,
> Toodih and El-Mikrat, whose trace is not yet effaced for all the spinning of the South winds and the northern blasts;
> …
> There, all about its yards, and away in the dry hollows you may see the dung of antelopes spattered like peppercorns …[13]

It is important to speak briefly to the hermeneutic structure of the ghazal as it ties directly into what inspired it: the desert (of life), and the Jāhiliyya poet. In the early period, the term *ghazal* denoted lyrical verse dedicated to love. It was not a formulaic and defined entity with codified, prosodical rules applied to a certain content form as in a sonnet but rather it consisted of 'thematic and stylistic conventions, and conventional imagery' ("Ghazal in Persian", *Encyclopedia of Islam*, 3rd ed.). It focused on affective intensity rather than originality, as it spoke to the transitory nature of life via the beloved. Images such as that of the desert and the remnants of a once-embracing beloved displaced by camel dung in the empty theater of decampment spoke to an oral pastoral tradition, although in time the purpose of such expression necessarily varied commensurate with the ecology of the poet:

> As a conceit (*maʿnā*) the early, desert-centered ghazal underwent centuries of metamorphosis and was adapted to other cultural-literary contexts. In twelfth-century Persia, for example, ghazal often represented a sense of existential anxiety in a hyper-intellectual and mystical register. Later, Urdu ghazals made love into the ultimate form of reason in response to a chaste love affair. Thematically, the notion of unrequited love in the poetics of the Arab poets differed markedly from that of their Persian counterparts as the Arab poet most often comes to terms with loss whereas the Persian does not. The evolution of the lost beloved from a physical and palpable being into a figure that enigmatically falls somewhere between real and divine highlights geopolitical changes and cultural interest in philosophical ideas and ideals.[14]

Existentially, the image of the mirage plays a crucial role within the larger spectrum of the desert. The *Jāhiliyya* poet experienced the mirage not as a forsaken nuance but as a hortative inspiration. When conceptualizing the language of this poet, one comes across a simple reflection of a far more complex conundrum as pertains to existentialism embodied in the out-of-reach nuance of the mirage against the all-encompassing backdrop of the desert. The hyperbolized—even fantastical—topological similarities as contained within the imaginary paradigm of an oasis and a mirage are nothing short of philosophical in terms of humankind's so-called aspirations and our wicked nature that prompts us to seek it. These epigrammatic poetic nuances are immersed in the all-consuming yellow dust of the desert and all who occupy it and that which occurs in it.

Every aspect of the bucolic life was somehow poignantly present for the *Jāhiliyya* poet by the desert and its components, and as such mores and values were esteemed, justified, and reprimanded by these said components:

The *Mu'allqāt* contain other glimpses of these more realistic mores and offer an alternative vision of pre-Islamic social ethic more compatible with those of Middle Eastern pastoral nomads. In the didactic movement that distinguishes Zuhayr b. Abī Sulma's (d. 609)[15] *mu'allaq* from the other classic odes, the poet cautions his audience:

> And he who doesn't conduct the bulk of his affairs with
> diplomacy and compromise; he'll be ground up by the camel's teeth and crushed by its hooves.[16]

As an ever-astute Boswell to desert's every whim and every act, the poet opted to personify its every aspect:

> There are traces of Kholah in the stony, sandy plain of Thahmad, which appear like the marks, (*lit. remains,*) of tattooing on the back of the hand.[17]

Desert, as such, takes on animate mannerisms whereby human action is therein determined. There are certainly many glimpses into this oral tradition whereby personification is presented as a useful source in creating florid metaphors—and extreme hyperbole:

> And the cloud poured out on the desert of Ghabeet its goods, (*i.e., rain*;) and it resembled the arrival of Yemani *mer chant* with his trunks loaded *with rich clothes*.[18]

We see here a brilliant use of three literary devices. There is antithetical congruity between the cloud (rain) and the desert; there is further congruity in the use of 'goods', i.e., rain and goods on the back of the camels traversing the desert. There is a *qarin* (here a hint) given as to the latter with 'trunks loaded with rich clothes'.

In an instance when Labid (d. 661)[19] has deemed himself the savior of his tribe, he says,

> And how many a chilly morning in which the reins of the cold had fallen into the hands of the frigid North Wind, have I [eased the people's suffering] with food.[20]

The desert determines *qazā va qadar* (fate). As in Persian poetics and the concept of *falak* therein, the desert presents itself as a worthy adversary—even an immovable object, a test of will, and a proprietor of anxieties that behoove the human condition. It is also an image well suited for Bedouin's view of humanity's relentless search for passion and love in the tumult of loneliness, i.e., water in the desert, life amongst the dead. At times, it seems also that the view of these poets' realities was but a mirage, commensurate with fascinations and hyperbolic views of self, tribe, and feats of existentialist rigor effectuated by the desert; again, another sense of the unreal real—as can be demonstrated in these two distichs by Amr b. Kulthum (d. 584)[21] and Imru al-Qays, respectively:

We cleave with them the heads of the people, and wo cut their necks, and they are cut.[22]

> She gives with thin fingers, which are not thick, as if they were the worms of the desert of Zabi, and soft as the tooth-brushes of the Isḥil tree.
>
> *(Johnson 2003: 16)*

These poets harness the minutiae (the micro-) in each particular desert. The constituents of each desert possess a certain poetic personality, even specific personified attributes, if you will, i.e., the desert of Zabi, the desert of Wajrah:

> She turns away, and shows me her smooth cheek, and is prohibiting me from caressing her with a glancing eye, like that of a wild animal, with young, in the desert of Wajrah.[23]

There is a resounding sound of a particular ethics at play in the works of most of these poets. The desert life had its vile beings, the thief, the murderer; however, somehow the poet finds cosmic logic in the trade of such folk as the poet, in a non-judgmental fashion, and sees pertinence between the trade and the nature of the said 'horror vacui', whereby human and wolf are the same, i.e., thief as the wolf:

> And many a valley like the plain of 'Aer, a sterile desert,
> have I crossed, in which the wolf was howling like the gambler with a family to support.

Morality, and lack thereof, we see, is often judged and punished under the rubric of the desert life—judged by the desert and executed by its otherwise usual suspects, in the first example, a camel:

> And he who doesn't conduct the bulk of his affairs with diplomacy and compromise; he'll be ground up by the camel's teeth and crushed by its hooves.[24]

It is interesting that the desert life and its harsh existential parameters bring out the best of the folk with mores and codes that define survival—not for one but for all. It is desert versus humankind, as Labid says[25] "I tell the [maysir players] to slaughter a she-camel, barren or pregnant, her meat given to all our neighbors' (Brown 1975: 39). The poet highlights how humanity is in direct competition with the animal for survival. He is even envious of the animal at times, and some of their expressions to that end highlight the desert's acute effect on behavior and speak to a far larger picture of humanity's ultimate inability to emancipate him/herself from nature:

> Seeking for a young calf thrown on the dust, of a whitish colour, whose limbs are dragged by grey hunting wolves, whose food supply is not cut off.[26]

Although we may see many examples of stylized and violent backlashes against other tribes in the poetry of the Jāhiliya, within their own communities, the poets voice comraderies that are uniquely democratic:

> And he who has great fortune and withholds it from his clan is cast aside and derided.... And he who gives their due to those who do not deserve it (*fi ghayri ahlihi*) their praise will be his demeaning and he'll regret it.[27]

As can be opined here, there is an alive, often unspoken, ethical pamphlet of accepted and unaccepted behaviors within each tribe. The concept of the desert as a void not only creates an ironic sense of unity for the sake of survival but also creates a prejudiced gap with the other tribes. This concept, which directly ties into the ecological nature of life for the Bedouin, rates a further explanation best summarized here:

> The celebration of life is entwined with a tragic sense of the inevitability of death and the infinite nature of life itself, and vice versa. This duality of positive and negative can be called the time of omnipresent tension.[28]

One's nature faces its greatest challenge in the desert. It is a domain that can constantly, and frequently, bring to the fore attributes otherwise less called upon: patience, bravery, gusto, etc. As such, the poet touches on humanity's lonely journey in life since that life is singularly personified in the vacuousness of the desert: mistrust, stoic demeanor, and introverted humanness all become part of what is prescribed. This did not imply enmity towards others, it implied survival of the self, best achieved by relying on oneself:

> And he who travels should consider his friend an enemy and he who does not respect himself will not be respected.

As Brown says, this could also mean 'And he, who is always seeking to bear the burdens of other people, and does not excuse himself *from it*, will one day by reason of his abasement, repent'.[29]

The environmental aspects of their poetics tie directly into the importance of, and their almost-human, connection to their beast, whether it be cows, donkeys, and very specially camels, the latter of which being in fact an extension of their existentialism, and symbolic of their existence. The animal, here the camel, which at times also provided milk and meat, in its perseverance along the arid path becomes a collaborator or sorts, a travel mate, a trusted 'only' other. Most importantly, the camel's uncanny ability to weather hardship served as a persistent reminder that survival is achieved by sheer will. This, along with the pastoral and nomadic significance of the animal within the nomad's existential paradigm, highlighted the camaraderie between man and beast, i.e., the irreversibility of this relationship. They also speak of human clutter that disturbs natural habitat and cosmic tranquility. One must not fail to see in all of these images, however, that it is the desert that nurtures such thoughts of tranquility, human existentialism, and symbols of undeterred fate:

> And the wild cows, having lately calved, are standing over their young giving suck, whose {i. e., the valley's) young lambs have herded in the plain.[30]
>
> Then I stood questioning them, concerning the inhabitants, And of what avail is our questioning rocks and boulders lying in their places for ever, whose speech is not clear?[31]

Although the Arab poet's *nasib*, the first 10–15 lines of his ode, was often dedicated to youth, love, and loss, one can see that the images used are not a far cry from what followed these lines, particularly as it pertained to the praise of warrior in the tribe, self-praise, etc. The respect for these personae was based on several factors, most important of which being that these were folk that lived the life of the desert and were of the same social fraternal code of conduct: all beloved by way of the desert. Also, it should be noted that these early depictions can't be equated to the empty post-Islamic praise of the patron for the sake of reward. Attributions were afforded not for mere lionizations which held no truth and were exaggerated attributions of a patron who most likely deserved none. Theirs was drenched with machismo and faithful fervor.

Perhaps the most important theme of the poet of this time is love. In the to and fro of images from the *nasib* to the *madh* (panegyric), the images were continuous as pertained to the praise of a beloved. The beloved of the desert is a known entity and the escapades of the poet (lover), and the beloved are subject to a philosophical enquiry which were hawkishly watched by this unknown: the desert, the universe. The identity of the beloved, in what follows the pre-Islamic poetry, was shaded in obscurity in the post-Islamic period in most places, particularly in the formalized ghazal and even in the *taghazzul*s (exordiums). Nonetheless, the same phenomenon applied: the language used to praise the patron, warrior poet, champion of the tribe, etc., used in the *madh* (eulogy), was merely the foundation for the imagery depicting the beloved. The only difference was that in the centuries that followed a palpable beloved was now an opaque beloved, who bordered on divine and represented an unrequited love, and like the patron this new beloved whose paradigmatic attributes were out of reach 'had been hyperbolized to the extremities of exaggeration'.[32] Therefore, the only opaqueness we experience in the poetics of the *Jahiliyya* is what the desert and its components have in store for two very much real beings.

Ultimately, the very crucial image of wine (*khamriyya*) is of interest to the poet as concerns their romantic interludes and/or remembrance of love lost. Although this elixir was used for different purposes at different times in the development of this image by different poets, for the *Jāhiliyya* it serves as a numbing agent against the harshness of the bucolic life and the details pertinent to their life within that theater. As any next station upon the desert is an unknown, so is any next station on the path trodden that reminds one of the beloved no more:

> A wine mixed with water and red as if mixed with *ḥuṣṣ*
> After drinking it, we will think no more of possessions,
> We will become beneficent, philanthropic.
> This wine will keep the needy free from desire,
> And all memories of their hardship will fade away.[33]

It is of importance to mention the significance of time and timelessness in the poetry of the Bedouin Arab of the pre-Islam period. Desert and its void afford two types of time—and movement—for the poet. When it came to panegyrics time stopped and so did movement for the poet of the desert. The poet captures the significance of one entity within the larger theater of the desert. There is almost no continuity, whereas in the didactic expressions, where we see a multi-layered narrative of *nasib*, etc., the poet speaks to the multi-faceted aspects of existentialism:

> time does not play an operative role in giving S.L. [single-layered structure] its semantic structure and semantic value. Contrary to this, M.L.S. [multi-layered structure] and M.D.S. [multi-dimensional stream] involve time as a fundamental experience, as a force shaping the consciousness of the poet and his vision of reality. Because of this, both types involve a process of narration, depicting entities in a time context of change and continuity (permanence), hence the description of atlal (effaced encampment) is a constituent unit of the multi-layered structure, although, owing to other factors related to the nature of M.L.S., the ruins do not appear in it frequently.[34]

The fleeting nature of life is embodied in the unknown continuum of the land. As such the peregrinations of the poet upon the desert become paradigmatic stations upon the stage of life where love is the ultimate unfound, missing integral: a mirage which embodies the noble harshness of the desert.

> Stop, oh my two friends, let us weep on account of the remembrance of my beloved, and her abode situated on the edge of a sandy desert between Dakhool and Howmal.

> And *between* Toozih and Maqrat, whose traces have not been obliterated, on account of what has blown and re-blown over them from the South wind and the North wind.[35]
>
> As if in the morning the small birds of the valley Jiwaa had taken a morning draught of old, pure, spiced wine [*mufalfal*].[36]

The word spiced *mufalfal* is not understood here by many scholars. It implies a biting taste, as in biting one's tongue; the double entendre here is that the term *mufalfal* also means a densely curly hair. This sense of intoxication connotates a compounding effect of being lost in the hair of a beloved while tasting her ruby (red wine) lips. As can be seen, the development of the theme is quite advanced for its time. All the while there is absolutely no attempt by the poet to emancipate himself from the habitat that inspires this romantic interlude: it's wild, erotic, and enchanting. In building these otherwise soft images, we see the development of a rigorously stern jagged edge inspired by the difficult desert path—always reminding one of a hindsight:

> As if in the evening the wild beasts in it drowned in the furthest parts of it, (i.e., the valley Jiwan,) were the root-bulbs of the wild onion.[37]

The infatuation with the desert may imply to some an otherworldly, even mystical desire to ignore all things otherwise deplorable or mundane, however, this is indeed far from the truth. It is a place of solace where mundane dreams are not disturbed by realities of unfulfilled dreams. Bravado doesn't stop short of bellicose situations and goes well beyond it. In fact, every aspect of desert life, nay life as we know it, at least as the poet in this era is concerned, is considered a brave act. The path is merely a constant which changes with environment and the peoples: life analogous to a camel, to a ship, to sea, to sand:

> As if the Malikian camels, with the howdahs on the morning of her departure on the water-tracts of the village of Pad, were the big ships of 'Adoul, or the vessels of Ibni Yamin, which the sailors at times steer out of the straight course, and at times guide straight.[38]

The poet has a panache for somehow comparing the tracking of a camel's footsteps to tracking a ship. Although indeed a ship leaves no tracks, it is not the track which is of concern for the poet, as track here implies a cosmic design in cahoots with the deserts and the sea and the sky: *qazā va qadar* (fate): "Their bows cleave the ripples of the sea, as the divider of the sand-heaps separates the dust with his hand'.[39]

From here the poet cleverly escapes into a description of the amorous beauty of a young girl. In describing this beauty, we see other images that help the poet brave the desert life, as does his camel, with its shining teeth, etc., creating a dichotomy between the beloved and the camel:

> And she is smiling with her deep red lips, and shows teeth like a jessamine blossoming in a damp sand-hill, situated in the midst of a plain of pure sand.[40]

In a brilliant image, the poet creates a cross-referential juxtaposition between jasmine in a damp sand hill and further highlights the nomadic adamance in finding life in the desert and in its otherwise arid deathlike infinitum. By exaggerating the depth of the red lips, a further palpability of the image is provided in relation to the desert and its depth. 'Damp' here takes us back to the fervent desire for possibility and hope in the midst of 'a plain of pure sand'.

The poet goes on to say:

> And as for me, verily, I banish my grief at the time it presents itself, by the help of a thin camel, swift in its paces, which travels by night and by day.[41]

The role of the beloved in the Arabic ode has been subject to a cross-generational morphosis. The diachronic-poetic nature of this phenomenon, in view of the onslaught of theosophical thought in Persian poetic lexicon, that uses these poets' poetic genius as a source, changed the image of the beloved and its previous simplicity and morphed it into a more oblique one: a divine *inamorata*. This new beloved shared a common purpose with a truly malevolent celestial force that had slowly developed its misanthropic literary reputation during the earlier centuries (10th–11th c.). With the beginning of the end of panegyric as a source of income, many poets in these centuries reflect their anger in the previously faultless beloved. This new coquettish beloved was now in the cahoots with the celestial sphere to shatter dreams by posing as a mirage of sorts. The *Jāhiliyya* poet, however, sees the beloved as an entity that experiences what they experience: they are both victims.

The beloved's dual nature as a betrayer of trust and an accomplice to the dark side of life on one hand and as the poet's chosen refuge on the other is a theme that begs analysis. However, what is for certain is that the beloved here does not have a mysterious collusion with the celestial sphere and has little resemblance to the *Jāhiliyya* beloved. This would demonstrate a functionality—via some of the more salient rhetorical techniques—that exists between this poetic morphology, in view of the *modus operandi* of the beloved in relation to the milieu, and subsequently the prevalent philosophies of the times, which eventually led to the creation of a rogue beloved forever where the beloved is inextricably entwined with any image that represents mystery and horror, i.e., the desert, their endless Celestial sphere.

Notes

1 For more on the many works on *Jāhiliyya* (age of ignorance) see William E. Shepard, "Age of Ignorance", in *Encyclopaedia of the Qur'ān*, ed. Johanna Pink (University of Freiburg). Consulted online on 03 August 2023 http://dx.doi.org/10.1163/1875-3922_q3_EQSIM_00013.

2 "The desert [also] evokes horror, nature as animated by plant and animal life inspires admiration, and both horror and admiration find constant expression in the works of the poets. But their comprehension of nature follows a different path from ours. The manner of expressing the relation to nature is undoubtedly beautiful, it abounds in splendour, but it is visually decorative, conventional, static, in the lyrics even stereotyped". For more see Jan Rypka, *History of Iranian Literature* (Reidel, 1968), 80.

3 Aidan Tynan, "Desert Earth: Geophilosophy and the Anthropocene", *Deleuze Studies* 10, no. 4 (November 2016): 489.
4 Reynold Nicholson, *A Literary History of the Arabs* (Cambridge University Press, 1966), Introduction.
5 Friedrich Nietzsche, *Thus Spoke Zarathustra: A Book for All and None*, trans. Adrian Del Caro (Cambridge: Cambridge University Press, 2006), 80 (as cited in Tynan, "Desert Earth", 482–483): "Nietzche touches on this notion: It was Nietzsche who first proposed that the philosopher emerged from desert ascetics: 'In the desert the truthful have always dwelled, the free spirits, as the rulers of the desert; but in the cities dwell the well-fed, famous wise men -the draft animals'".
6 Herbert N. Schneidau, *Sacred Discontent: The Bible and Western Tradition* (Berkley: University of California Press, 1976), 143 (as cited in Tynan, "Desert Earth", 482).
7 "A name given them at a much later date by Ḥammād-al Rāiya, and on which is founded the absolutely untrue legend that these pieces, written with golden ink, were hung up in the famous Temple of Mecca, in the Ka'ba. The name was merely intended as an allusion to the place of honour they hold on to the Arabian Parnassus, even as a chandelier may be suspended in the midst of any apartment, of rather as a necklace may be worn hanging about the neck—for they were also called *al-Summūṭ*, "the necklace of pearls." See Huart, *A Hsitory of Arabic Literature*, 10. As true as this estimation may be of the origination of the term 'hanging' the author of Huart does not take into consideration the importance of the desert night and the arrangement of the stars as possible inspirations for the term; i.e., the hanging of the skies upon humanity, which in turn inspired the notion for the term *manẓūm* 'arranged' in reference to poetry.
8 Khāqāni Shirvāni, *Dīvān-i Khāqāni Shirvāni*, ed. Mir Jalāl al-Din Kazzāzi (Nashr-i Markaz, 1996).
9 F. E. Johnson, *The Seven Poems, Suspended in the Temple at Mecca* (Education Society's Steam Press, 1893), 101.
10 Huart, *A History of Arabic Literarture*, 40.
11 Like the *qasida* a mono-rhythmic poem. It literally means fragmented or piece. It is much shorter in length and does not possess the thematic layers of the *qasida* mentioned earlier.
12 He is considered the father of Arabic poetry. The distich quoted here comes from his famous *qasida* "Let us hald and shed tears" (qifā nabki). It is one of the seven Mu'allaqat, poems prized as the best examples of pre-Islamic Arabian verse. Imru' al-Qais was born in Norther Arabia iun the Al-Qāsim region. He was also the king of the Kindite kingdom (450–550) of the Ma'add federation of tribes in Northern Arabia: "And he is also called [...] (the much-erring king), on account of his amorous tendencies. He fell in love with 'Unaizah, the daughter of his uncle Sherhabeel, and of these two lovers there is a tale which the poet tells in the poem". For more see Johnson, *The Seven Poems*, 16.
13 A. J. Arberry, *Seven Odes: The First Chapter in Arabic Literature* (The MacMillan Company, 1957).
14 Alireza Korangy, "Love and Reason in the Ghazal", in *A Companion to World Literature*, ed. Ken Seigneurie (Wiley, 2020).
15 A *Jāhiliyya* poet from the Banu Muzaina tribe whose poetry truly engages the desert existence: pillage, plunder, the magnificence of the desert, and boasting of one's tribe.
16 Jonathan A. C. Brown "The Social Context of Pre-Islamic Poetics: Poetic Imagery and Social Reality in the Muᶜallaqat," *Arab Studies Quarterly* 25, no. 3 (2003): 40.
17 Johnson, *The Seven Poems*, 31.
18 Johnson, *The Seven Poems*, 29.
19 Abū Aqīl Labīd b. Rabīʿa b. Mālik alʿĀmirī, a warrior of the tribe of Hawazin, who late in his life became a Muslim. Only one complete *qasida* of his is included in the *Mu'allaqāt*.
20 Johnson, *The Seven Poems*, 47.
21 "He is said to have recited this poem extemporarily in the presence of the King, 'Amru, son of Hind or Munzir, on the occasion of his bringing about peace between the two

tribes of Bakr and Taghlib, who were on the point of renewing their war. The poem is said to be a long one, containing about 1,000 lines, the poet having added a good deal subsequently… The poem coatains a lively description of the chivalrous deeds of the tribe of Taghlib and their glories". For more see Johnson, *The Seven Poems*, 128.

22 Johnson, *The Seven Poems*, 143.
23 Johnson, *The Seven Poems*, 14.
24 A. C. Brown "The Social Context of Pre-Islamic Poetics", 39.
25 Robert Dirks, et al. "Social Responses During Severe Food Shortages and Famine [and Comments and Reply]," *Current Anthropology* 21, no. 1 (1980): 21–44: "The anthropological study on the effects of food deprivation and famine on societies, however, has faced the practical and moral difficulties inherent in the subject; in such times of hardship, it is difficult to isolate social science variables and impossible to refrain from intervening with support. Nonetheless researchers have documented various societal reactions to both perennial scarcity and years of extreme famine".
26 Johnson, *The Seven Poems*, 105.
27 C. Brown "The Social Context of Pre-Islamic Poetics", 42.
28 Kamal Abu-Deeb, "Towards a Structural Analysis of Pre-Islamic Poetry," *International Journal of Middle East Studies* 6, no. 2 (1975): 151.
29 Johnson, *The Seven Poems*, 87.
30 Johnson, *The Seven Poems*, 93.
31 Johnson, *The Seven Poems*, 22.
32 Alireza Korangy, Development of the Ghazal and Khāqānī's Contribution: A Study of the Development of Ghazal and a Literary Exegesis of a 12th c. Poetic Harbinger (Harrassowitz, 2013).
33 Ahmad Torjānizada, *Sharh-i Muʿallaqāt Sab*ʿ (Sorush, 2003).
34 Deeb, "Towards a Structural Analysis of Pre-Islamic Poetry", 152.
35 Johnson, *The Seven Poems*, 2.
36 Johnson, *The Seven Poems*, 29.
37 Johnson, *The Seven Poems*, 30.
38 Johnson, *The Seven Poems*, 32.
39 "Some small article, such as coin or ring, is buried in a heap of sand, the players all staking similar amounts. The heap of sand is then divided by one of the players, (called the مفايل) into a number of smaller heaps—one for each player—the player in whose heap the article is found wins the stakes." See Johnson, *The Seven Poems*, 33.
40 Johnson, *The Seven Poems*, 34.
41 Johnson, *The Seven Poems*, 35.

Bibliography

Abu-Deeb, Kamal. "Towards a Structural Analysis of Pre-Islamic Poetry." *International Journal of Middle East Studies* 6, no. 2 (1975): 151.

Arberry, A. J. *Seven Odes: The First Chapter in Arabic Literature*. The MacMillan Company, 1957.

Brown, Jonathan A. C. "The Social Context of Pre-Islamic Poetics: Poetic Imagery and Social Reality in the Muʿallaqat." *Arab Studies Quarterly* 25, no. 3 (2003): 29–50.

Dirks, Robert, et al. "Social Responses during Severe Food Shortages and Famine [and Comments and Reply]." *Current Anthropology* 21, no. 1 (1980): 21–44.

Huart, Clement. *A History of Arabic Literature*. Appleton, 1903.

Johnson, F. E. *The Seven Poems, Suspended in the Temple at Mecca*. Education Society's Steam Press, 1893.

Khāqāni Shirvāni, *Divān-i Khāqāni Shirvāni*. Edited by Mir Jalāl al-Din Kazzāzi. Tehran: Nashr i Markaz, 1996.

Korangy, Alireza. *Development of the Ghazal and Khāqānī's Contribution: A Study of the Development of Ghazal and a Literary Exegesis of a 12th c. Poetic Harbinger*. Harrassowitz, 2013.

Korangy, Alireza. "Love and Reason in the Ghazal." In *A Companion to World Literature*, edited by Ken Seigneurie, 1–12. Wiley, 2020.

Nicholson, Reynold C. *A Literary History of the Arabs*. Cambridge University Press, 1966.

Nietzsche, Friedrich. *Thus Spoke Zarathustra: A Book for All and None*, trans. Adrian Del Caro. Cambridge University Press, 2006.

Rypka, Jan. *History of Iranian Literature*. Reidel, 1968, 80.

Schneidau, Herbert N. *Sacred Discontent: The Bible and Western Tradition.* University of California Press, 1976.

Shepard, William E. "Age of Ignorance." In *Encyclopaedia of the Qur'ān*, edited by Johanna Pink. University of Freiburg. Consulted online on 03 August 2023. http://dx.doi.org/10.1163/1875-3922_q3_EQSIM_00013.

Torjānizada, Ahmad. *Sharh-i Mu'allaqāt Sab'*.Sorush, 2003.

Tynan, Aidan. "Desert Earth: Geophilosophy and the Anthropocene." *Deleuze Studies* 10, no. 4 (November 2016): 479–495.

7

EXPLORING *'Y PAITH'*

Historical and Contemporary Perspectives on Desert Environments in Patagonia, Argentina

Hywel M. Griffiths, Stephen Tooth and Gabriel Kaless

> *paith*
> a: desert, wilderness, uninhabited land; devastation, ravage; also fig.; deserted, uninhabited, ravaged.
> b: scene, prospect.
> c: treeless plain(s), esp. in Patagonia, prairie, pampas.[1]

Introduction

Hywel: Thinking of Wales in the Patagonian Desert

I watched small rivulets of water weave over the grey-brown earth, and listened to the turbulent streams, flowing as though determined to enjoy this brief, rare event. The scene around us was treeless for the most part, but our paths along the riverbanks meandered between the thorny shrubs and bushes that grew close together, adding a muted, light green hue between soil and leaden sky. I had been looking forward to being in this place after three days of workshops at La Universidad National de La Patagonia San Juan Bosco in Trelew and El Centro Nacional Patagónico (CONICET-CENPAT) in Puerto Madryn (Figure 7.1), with their focus on flooding, drought and ecosystem services of desert rivers, floodplains and wetlands. The workshops had mixed audiences of students, academics, representatives of governmental and other decision-making authorities, tour operators, and local community members and demonstrated how scientific research can be combined with other information sources (e.g. historical documentary records, community memories, personal experiences) to address pressing environmental, economic and social issues, such as mitigating flood hazards, evaluating catchment management strategies, and developing geotourism ventures. To complement the more

DOI: 10.4324/9781003397212-10

technical discussions about flood monitoring, I had eagerly anticipated being out 'in the field' with Gabriel and his colleague Andrés, and with Stephen and particularly to visiting the desert ephemeral river catchments[2] instrumented with time-lapse cameras, flow gauges and sediment samplers, all located in tributary catchments upstream of a reach of the lower Río Chubut (Afon Camwy in Welsh) called Boca Toma (Ceg y Ffos in Welsh). In the town of Gaiman where we were staying (Figure 7.1), and in many other areas of *y Dyffryn* (the valley, specifically the lower Afon Camwy valley), the electricity had not been restored following the previous night's intense and dramatic thunderstorm, and a planned interview to local media about our workshops was cancelled. But, with more time to explore, we made for the tributary catchments in Gabriel's pickup truck. We crossed the lush, green, irrigated fields of the lower Chubut floodplain, then drove past the village of 28 de Julio, before rising onto the greys, light browns and softer greens of *y paith.*

Just over five years previously, in 2014, Stephen and I had made our first visit to Welsh Patagonia, also known as *y Wladfa* (literally, the colony) in Welsh. We had driven from *y Dyffryn* to the Welsh communities in the Andes (Figure 7.1), a journey of some 680 km, in a small car. On our two-day journey, we passed only a few travellers going the other way on the tarred highway that we followed to just beyond Los Altares (Figure 7.2A), and fewer still on the dirt roads that followed the river upstream to Piedra Parada. This 2014 visit had taken place in the blazing sunshine of the southern hemisphere's summer. During our 2019 visit, the

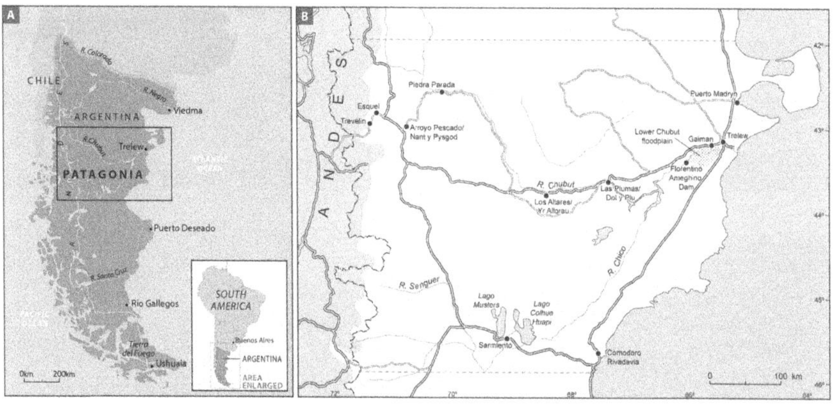

FIGURE 7.1 Location of Patagonia, a loosely defined region (darker shade on left-hand map) that stretches north from Tierra del Fuego to the Río Colorado and incorporates areas of present-day Chile and Argentina. Patagonia includes desert areas, mainly east of the Andes in Argentina. The main map shows some of the key geographical features, present-day settlements, and main roads in Welsh Patagonia, another loosely defined area largely corresponding to Chubut Province.

rain and relative cold of the southern hemisphere's early autumn made y *paith* feel very different. The dirt road that we followed crossed a dry riverbed before leading us to the first instrumented site. The gauge, installed to measure river flow and sediment transport during flood events, and monitored by a time-lapse camera on the riverbank, stood in the middle of the dry riverbed (Figure 7.2B). We drove on to explore other gauges, walking along riverbanks and crossing hillsides. We heard more thunder, and it continued to rain, seemingly ever more heavily.

On returning to our original crossing point, we found that the formerly dry riverbed was now a fast-flowing stream (Figure 7.2C). We resolved to return later, when surely the levels would have dropped. In the meantime, we drove to other tributary catchments and observed with wonder and with camera shutters clicking, the rare desert flash flood event.

We returned to the original crossing point twice, and each time had to turn back due to water levels that seemed still to be rising, despite the thunderstorm having abated. We were tired, wet, and wondering how we would get back to *y Dyffryn*. Watching the small, innumerable rivulets and hillside streams, all swelling the main rivers (Figure 7.2D), my thoughts turned to home, family, and the mid-Wales uplands (called by some, ironically, the 'desert of Wales' due to a perceived lack of settlement, infrastructure, and biodiversity[3]). Beyond these particular uplands,

FIGURE 7.2 Views of *y paith*: (A) in summer (January 2014); (B) in autumn (March 2019), with a gauging station (white tube on lower right) located in a dry desert riverbed; (C) same location a few hours later, with fast-flowing water now almost at bank tops; (D) runoff arising in rivulets and small hillside streams continuing to swell flow in the main rivers.

the term 'desert' or Welsh equivalents also have been used metaphorically, usually in a negative sense, for example, in Idris Davies's poem Gwalia Deserta,[4] which depicts impacts of capitalism on south Wales's industrial areas in the early twentieth century. This sense of strangeness and displacement, but also familiarity, was in our minds. How long would we be on *y paith*? Where would we stay? How much would it cost to re-arrange our travel plans and get home? No doubt numerous desert travellers before us had been faced with similar dilemmas, for many deserts worldwide are subject to hydrological extremes (flood and drought) that have posed challenges for those unfamiliar with such environments.[5] Indeed, on *y paith* in the Afon Camwy catchment, a Welshman, thinking longingly of familiar Welsh landscapes in an unfamiliar desert landscape, his movements dictated by its hydrological extremes, was not a new occurrence.

Background

The Welsh colony of Patagonia (*y Wladfa*) was established in 1865 by around 153 adults and children who had sailed on board the tea-clipper, *Mimosa*, from Liverpool, England, to *y Bae Newydd* (the New Bay, later the site of Porth Madryn/Puerto Madryn – Figure 7.1) on Argentina's Atlantic coast. Shortly after arrival, the party relocated to *y Dyffryn*, and this became the early colony's focal point. The emigration was not unique; during this period, Welsh citizens had emigrated to Australia, Brazil, New Zealand, Canada, South Africa, and in particular, North America.[6] Reasons for emigration were varied but included escape from limited economic opportunities available in traditional agriculture or arduous conditions in the burgeoning heavy industries (especially coal and slate mining), as well as a search for new opportunities elsewhere.[7] In newly settled regions, however, Welsh culture and language were usually gradually lost as settlers assimilated successfully into the typically wider anglicised cultures.[8] So alongside any economic opportunities, emigration to distant, isolated Patagonia was also an attempt to settle somewhere where this cultural dilution and assimilation could be avoided; the colonists could experience economic freedom while practicing their religion and culture in their native language.[9]

The relationship of the Welsh colonists with the nomadic Indigenous peoples who lived in Patagonia – the Tehuelche, Mapuche and Pampa – was complex. Although ultimately the Welsh were settler colonisers, paradoxically this occurred from a colonised position[10] because they "felt their nation to be stripped of sovereignty, socially disparaged, culturally quashed'.[11] Initially, there were peaceable relations: the Indigenous peoples helped the Welsh colonists to adapt to the unfamiliar environment, and there was trade and exchange of goods. The Welsh colony was, however, supported by the Argentinian state.[12] During the Conquest of the Desert in the 1870s and 1880s – a military campaign directed mainly by General Julio Argentino Roca that led to genocide of Indigenous groups – the Welsh colonists did attempt to intercede on the Indigenous peoples' behalf, but in a way that reflected colonial power structures.[13]

The remainder of this chapter explores historical perceptions of *y paith*, as recorded in a selection of archival material. This is drawn from Welsh-language materials, which are some of the primary sources associated with the Welsh colony. Indigenous archival sources and voices from the nineteenth century tend to focus on the impacts of the Argentine military and engagements with the Argentine state, rather than on engagements with Welsh colonists.[14] The privileging of European voices from the archive in this history of Welsh Patagonia likely contributed to what Lublin has described as the enduring Myth of Whiteness in Argentina and the prominence of the European creation myth, despite decades of Indigenous activism.[15]

Following archival work on historical environmental perceptions in Welsh Patagonia and other global locations,[16] the Patagonian material analysed here includes colonists' letters and memoirs, and poetry by two key Welsh Patagonian poets, namely R. Bryn Williams (1902–1981) and Irma Hughes de Jones (1918–2003). Taking inspiration from the creative (re)turn in geography,[17] we weave our shared, yet very personal desert stories as a geographer and poet (Hywel), hydraulic engineer (Gabriel), and geomorphologist (Stephen). We reflect on visiting and/or living in desert environments, including further commentary on our experience of the March 2019 flash flood.

The Desert in the Archive

Alongside Michael D. Jones, Hugh Hughes (also known as Hughes Cadfan) was one of the principal architects of the Welsh emigration and had described the Patagonian landscape and climate in his *Llawlyfr y Wladychfa Gymreig* (*The Handbook of the Welsh Colony*), published in 1862. During his 1832–1834 visit to eastern Patagonia with *The Beagle*, Charles Darwin had described the landscape as having a "very sterile appearance; it is covered with thorny bushes and a dry looking grass and will for ever remain nearly useless to mankind" and stated that it was "pronounced by all to be most wretched and useless".[18] In Hughes's handbook, however, Patagonia is described as having luscious pastures and receiving 142 rainy days annually[19]; a figure calculated through interpolation from scant temperature and precipitation data in areas far to the north and south.[20] Whereas Hughes's interpolations were likely intended to portray the environment in terms that would appeal to potential colonists, and may fortuitously have resembled true values in parts of Patagonia lying in the Andean foothills, it was an inaccurate representation of the eastern coast and the low-relief, more arid terrain surrounding the Afon Camwy valley.[21]

Coming from a green, relatively equable, and often rainy Wales, the colonists adapted rapidly to the harsh, capricious, and unfamiliar hydrological environment. In the first few years, the colonists tried to nurture crops in a drought-prone environment or protect them from the large flood events which periodically beset the colony, but often in vain. The colonists' fears and frustrations were highlighted in some of their first reactions in letters sent to friends and family in Wales. Immediate reactions of the first colonists were commonly related to the lack of water, and comparing their new surroundings to home. William Jones expresses in a letter to his family on 7th November 1865, just over three months after disembarking:

> There is no fresh water here, except for that which stays as pools on the land after rain, that which I have not had any more of since being here than I had in one day in Bala many times.[22]

Similar to the experiences of other European colonists in deserts worldwide,[23] Jones is reflecting on the unpredictable nature of rainfall and associated surface water resources, a considerable threat to early colonists' survival.

Later, disruptive floods were the main concern of colonists in *y Dyffryn*. A large flood occurred in late 1865 and flood disruption occurred again in 1868, 1869, 1870, 1879, 1880 and 1884.[24] A particularly large flood in 1899 destroyed many parts of early settlements in *y Dyffryn* and forced some of the inhabitants to flee to hillsides to live in tents. Significant floods also followed in 1901, 1902 and 1904, and again in 1932 and 1958.[25] Even as early as the 1880s, with the continued flood disruption and land in *y Dyffryn* starting to become scarce, some of the colonists started looking and exploring westward over *y paith*, dissected by the Afon Camwy and its tributaries (Figure 7.1), toward the Andes.

These early encounters with the Patagonian desert were instrumental in the use of *y paith* in Welsh writings. Early authors avoided using the words *anial* or *anialwch* (both of which mean desert, with negative connotations), and preferred instead to use *paith* which could be translated into Spanish as *pradera*, and latterly has been translated as *campo, meseta, campaña, desierto, estepa*, or *pampa.*[26] Sarmiento's (1845) framing of *y paith* as a desert in the nineteenth century enabled depictions of the landscape as 'savage': an empty landscape in need of occupation and civilising.[27] Fernando Williams argues that the Welsh colonists saw *y paith* in the context of the Holy Bible, especially the Old Testament, and that they built their imagined relationship with the desert through this lens.[28] Patagonia was thus seen as the desert of the Old Testament, with the Welsh likening themselves to the Israelites in search of the promised land.[29] This perception has many similarities with the responses of colonists in other deserts, including parts of the American Southwest.[30] Further, the colonists' adversity and suffering in their inhospitable environment was seen in the context of Puritanism as identifying the colonists with God's chosen people. The desert embodied adversity, and some colonists took the view that the more hostile the conditions, the greater the dignity of those who decided to defy that hostility.[31]

The memoirs of John Daniel Evans – nicknamed *El Baqueano* (the leader) for his role in explorations westward from *y Dyffryn* – were published in a biography[32] and reveal the agential role of desert rivers in facilitating expeditions across the harsh *paith*, and ultimately the establishment of settlements such as

Trevelin and Esquel in the Andean foothills (Figure 7.1). The main desert river valleys – originating in the distant Andes and fed in part by snowmelt – were obvious exploration routes. Travel was easier along relatively flat floodplains, typically there was water year-round for the explorers and their horses, and fish and other wildlife provided food supplies. The descriptions in Evans's memoirs regularly note that they are following the Afon Camwy, or the Afon Fach (Río Chico), one of its major tributaries (Figure 7.1). The large woody debris transported by the rivers and deposited in channels or on the banks and floodplains also facilitated these expeditions, as they provided firewood and wood for repairing wagons.[33] At times, however, rivers also hindered expeditions, with Evans providing accounts of having to improvise in order to cross rivers without the aid of bridges or fords.[34] As in other desert environments worldwide,[35] we can see that the desert rivers provided affordances and hindrances to the colonial process, and that their hydrological, ecological and geomorphological characteristics were central to the colonists' evolving sense of the geographies of these 'new lands'.

The exploration and eventual settlement of *y Dyffryn*, *y paith* and the Andean region was also closely tied to river geomorphology in other ways. For instance, the *drofa* (turn, or meander in the river) likely would be sought as places to make camp and is an enduring feature of *y Dyffryn's* toponomy (e.g. as recorded in farm names such as Drofa Dulog). In later years, many settlements across *y paith* were located close to the scattered desert wetlands known in Spanish as *mallines* (i.e. floodplain wetlands, or wet meadows) where water supplies and livestock grazing are more reliable.

How do these historical perspectives regarding *y paith*, known at least in part by some present-day residents and visitors to Welsh Patagonia but at risk of fading with time, permeate and influence the present? Below, we offer some personal reflections, based on our varied experiences of *y paith*, including the flash flood in March 2019.

Hywel: Geography and Poetry in *y Paith*

My own poetry, inspired by the 2014 visit,[36] was strongly influenced by the general perception of *y paith* created in poetry, prose, film and other art forms and explored similar themes as some earlier poets, including memory of landscape and distinctive environmental processes (rain, flood, drought, and sedimentation). Even having experienced the contemporary landscape, it was difficult to avoid employing somewhat stereotypical imagery of *y paith* – isolated, lonely, inhospitable – due to the degree to which these perceptions are so deeply embedded in someone who does not live in Patagonia.

Indeed, my experience perhaps reveals how external perceptions of *y paith* may become entrenched over time through these creative practices.[37] In 1964, the main Welsh language cultural festival – the National Eisteddfod – was held in Swansea, south Wales. In anticipation of the centenary of *y Wladfa* in 1965, the topic of the chair competition – one of the two main prizes for poetry awarded annually by the Eisteddfod – was offered for an ode to 'Patagonia'.[38] The winner was R. Bryn Williams. Born in Blaenau Ffestiniog in 1902, his family moved to Patagonia when he was seven years old. Williams then returned to Wales in 1923 and became a historian, poet and playwright. He won the chair for a second time in 1968 and served as the *Archdderwydd* (arch-druid) of Wales – the main ceremonial office associated with the Eisteddfod – between 1975 and 1978. One of Wales's foremost writers on *y Wladfa* in fiction, non-fiction and poetry, Williams was, and posthumously continues to be, an important cultural figure. His writings were published around the time of *y Wladfa's* centenary, which itself inspired a new wave of interest in the colony's history,[39] and were influential in creating a perception of the Patagonian landscape in Welsh society during a period when most Welsh citizens would not have had opportunities to travel to Patagonia.

His ode of 322 lines, composed in the ancient Welsh strict metre form of poetry called *cynghanedd*,[40] is comprised of five sections, parts of which are provided below, translated into English by Hywel. In the first stanza, the poet hears the voice of Patagonia in the harsh wind, calling him back to *y paith*[41]:

Utgorn dy wynt gerwin, dig,
Taer heno o'r tir unig,
Mae'n galw, a'm hannog eilwaith
Yn ôl i bell anial baith.'

'Your angry, harsh wind is a trumpet,
tonight, earnestly, from the lonely land
calling, urging again
to return to the far, desolate *paith*.

In the stanzas that follow, we hear the voice of prehistoric Patagonia, and then the Spanish Christians arrive in an 'empty', 'burnt', and 'cruel' landscape[42]:

Y paith's furnace in the Valley of y Peithion[43]

The Spanish Christians, according to the poet, were the first 'pioneers' to 'tame the wild valleys'.[44] The next section describes this conquest's impact on the Indigenous communities, seen through the eyes of the last 'chief'. The Indigenous people 'lived free between mountain and sea' until the arrival of the '*Cristianos*' and their weapons, whereafter they were imprisoned 'like powerless sheep'.[45] While recognising the wrongs done to the Indigenous peoples, a paternalistic attitude towards these peoples is evident, which creates a misleading image of their disappearance from the region and their future fate. The section ends:

Knowing that hope will not return
Is the pain of the Chief of *y Paith.*[46]

The third section describes the arrival of the Welsh colonists. Their progress across *y paith* is portrayed as a heroic venture against the odds and underlines the perception of *y paith* as being desolate and unforgiving:

> Venturing, taking steps without stopping
> With the soul's confidence over the desolate desert-land.
>
> From valley to valley with the fickle winds,
> shivering, yet vanquishing in the soft gravels;
> because, with miracles supporting them – they journeyed
> with fervent glee and trudged persistently.[47]

The desert landscape is further portrayed as being inhospitable owing to heat and lack of rainfall. These climatic factors lead to crop failure and the suggestion of salinisation issues (see 'white dust' in the following excerpt):

> The ancient sun, from the wide sky
> on the land with its fiery rays,
> frail shoots withering,
> it scorched the crops of the land with its heat:
> The parched ground was burnt to a crisp,
> the white dust was the drift of woe:
> prayers were in vain,
> the rain did not come to the locked land.[48]

The ode memorialises the efforts of the colonists to adapt to this inhospitable environment. The irrigation work of the colonists is seen as 'civilized': application of the flow of the Río Chubut/Afon Camwy turns 'acres of dust' into a 'green country', 'like a broad garden'.[49] The first attempts to irrigate the land by making gaps in the river banks are memorialised:

> Where there is anguish, a better world will come,
> man's civilised skills and salvation:
> in their desert,
> turning the flow to the acres of dust:
>
> from the gaps in the river
> to the grey land the flow ran:
> from hill to hill, from the irrigation
> came hay and corn;
> and Camwy now everywhere,
> a green country, like a broad garden.[50]

This was painful work, requiring sacrifice and courage, which then led to a sense of ownership over *y paith*:

In their desert
they reaped the full grain from the dust.

From their sacrifice came our hope,
Their pain gave us *y paith*.[51]
The brave ones on the bare desert-lands
challenged death and suffered anguish.[52]

The colonists, having adapted to this 'new land on *y paith*', are joyful and have found a place where they can live the free life that inspired their original migration:

Young, we turn to the new land – on *y paith*,
Despite the hot winds of the summer days;
With joyful song, full of progress;
Never fearing to live freely.[53]

When the poet praises the colonists' mutual hospitality – 'they gave a cordial in a desert/to the feeble wretches of thorns and dust'[54] – it is easy to imagine this hospitality as a metaphor for turning the waters of the Afon Camwy to land irrigation.

The final section closes the ode with the poet pledging to return to Patagonia from Wales. As this ode won the National Eisteddfod chair, the image of the Patagonian environment and the adaptation of the colonists has become a key way in which many Welsh speakers imagine Patagonia.

A different perspective is offered by one of the other main Welsh Patagonian literary voices of the second half of the twentieth century. Irma Hughes de Jones was born on Erw Fair farm in Treorci in *y Dyffryn* and spent much of her life there.[55] Aspects of her poetry also reflect on the adaptations of the first few generations of colonists to the desert environment, after completion of the irrigation canal network and the Florentino Ameghino dam. In her poem '*Y Sul cyntaf* (the First Sunday) (Trerawson, 1965)', she imagines the early colonists' tears being dried by the sun, and the fact that they had left the dismal rain of Wales behind was a consolation in the desert.[56]

In 'Camwy', she imagines the 'Chupat' (the Tehuelche name for the Río Chubut/Afon Camwy) before the establishment of the colony, the irrigation canals, and the Florentino Ameghino dam. Her use of the old name Chupat is deliberate, as she sees the river as one that used to flood in a passionate and turbulent way, before the irrigation canals enabled the colony to experience the river's 'incredible virtue' in giving them corn and fruit from the earth, and slaking their 'long thirst'.[57] She recognises that after each flood, the quietness returned, and the 'fertile sediment' remained to enrich the land.

She finds happiness in how the wealth that the colonists had brought to the land of their desire remains, and compares it to the 'fertile sediment' that the 'grey waters' left behind on their way to the sea:

'Ni chofir bellach
am aberth, am wyrth yr arloesi,
ond ar ddyddiau gŵyl...
...Ond erys eto ryw 'ddau neu dri'
sy'n para'n ffyddlon...
...daw eco'r gân i gadw'n fyw rhyw atgof
am erydr gloywon....
...A'r llawenydd o weld aros
o'r golud a gludwyd
gan gynhyrfus eneidiau
i wlad eu dyhead,
eu Camwy hoff,
fel y gwaddod ffrwythlon
a edy'r dyfroedd llwydion
ar eu taith tua'r môr.'

'There is no memory now
of the sacrifice, of the miracle of the innovation,
except on festival days...
....Yet again, one or two are left
who remain faithful...
...the echo of the song keeps some memory
about the shining ploughs alive...
...And the joy of seeing
the wealth that had been transported
by the excited souls
to the land of their desire,
their dear Camwy,
like the fertile sediment
that the grey waters leave
on their way to the sea.'

To me as a geographer and poet, these poems reflect on memory and forgetting in the context of adaptation to the desert landscape, and also speak to a lay geomorphological knowledge borne of living and working on the Afon Camwy floodplain. Poetry has woven this knowledge into the stories that are recorded about the arid environment of *y Dyffryn*.

Gabriel: Engineering in *y Paith*

It often seems as if nothing happens on the endless Patagonian desert plains. Life is scarce and time is meaningless as the landscape is always the same. These feelings echo the thoughts of Sarmiento, the nineteenth-century president of Argentina and author of '*Facundo: Or, Civilization and Barbarism*' in which the Argentinian landscape is empty; a land with no owners, and as such, a land to be occupied.[58] But this is not the truth, of course. Inhabitants have been here for thousands of years. Moreover, *y paith* sometimes knocks on the door.

I have lived in Patagonia for almost half a century; first, in wetter Tierra del Fuego Island, and later in the more arid lower Afon Camwy valley (*y Dyffryn*) (Figure 7.1). The Patagonian landscape is a part of me and because of that, and my job's everyday duties, I seldom forget I now live in a desert. The desert knocks on my door intermittently and in different ways: flash floods, droughts, and turbidity events – when the river water is full of sediment washed from the sparsely vegetated hillslopes – are all reminders. These events have one common element: excess, whether expressed in terms of abundance or absence. Desert life is learning to adapt to nature's excesses.

Floods

Flooding of *y Dyffryn* has been of concern since the first Welsh colonists' arrival. The valley flooded the year they arrived and blighted early attempts to grow crops. Defences such as longitudinal levees were employed, and river straightening and diversion were also discussed but not implemented. For a while these measures worked, but proved insufficient to contain the 1899 flood, the so-called 'great flood'[59]. The valley turned into a lake in just a week and only the city of Trelew, built on higher ground, survived. The great flood was followed by smaller floods in 1901, 1902 and 1904. The situation required a major engineering intervention: the Florentino Ameghino dam, finished in 1968, was built across a narrow bedrock canyon some 100 km upstream of Trelew. Since then, Afon Camwy's waters appear to have been tamed. Some historical floods were forgotten: they belonged to the past, and new land became available for urban expansion.

But the water continued to arrive, this time as intense rainfall causing flash floods on the slopes around *y Dyffryn*. The low-lying land, now occupied by new neighbourhoods, was the worst damaged and people had to be evacuated from Trelew in 1976 and 1998. Further measures were implemented: urban drainage systems were designed and included earth dams, channels, transitory storage reservoirs, and pumping stations to return water to the river. After dam construction, river flow was more uniformly distributed throughout the year, and flow peaks did not reach the former dangerously high levels. The river channel that previously had to accommodate these high flood peaks started to narrow and the riverbed began to silt up due to vegetation colonisation and bank stabilisation. The channel at the Poet Bridge at Rawson city (near the river mouth), narrowed from 120 m to 40 m, reducing its conveyance capacity and, ironically, increased flood risk.

Drought

Most of desert Patagonia lies in the shadows of the Andes. Precipitation is scarce and losses through soil evaporation or plant transpiration (collectively termed evapotranspiration) are high (annual potential evapotranspiration increases south to north from 700 mm to 1400 mm). The Welsh colonists soon realised that crops would not thrive without additional water supplies. Abandoned palaeochannels of the Afon Camwy were initially used as irrigation canals. By 1883 most of the valley was irrigated, and irrigation companies were created to manage the resource. The Florentino Ameghino dam guaranteed water supply, or so it was thought. The Afon Camwy experiences alternating wet-dry cycles, each approximately ten years long, that are related to the ocean-atmosphere system known as the El Niño Southern Oscillation. In 2022/2023, we hope we are at the end of a dry period. The water surface level behind Florentino Ameghino dam is alarmingly low and water has been supplied with great care. Downstream conflicts have arisen between water users: the irrigation company (that diverts discharge into the canals) and fresh water supply cooperatives (that provide water for the population) face problems in pumping river water.

Turbidity

In 2017 the desert disrupted the life of *y Dyffryn* again, but in an unexpected way. Similar to the March 2019 event, heavy rain fell over the city of Comodoro Rivadavia (Figure 7.1) in March 2017 and caused catastrophic damage. The event seemed, however, to be far away from the people of *y Dyffryn*. However, the event also impacted the Afon Fach/Río Chico, an otherwise rarely activated tributary of the Afon Camwy (Figure 7.1). Heavy rain and scarce vegetation cover led to soil erosion. Sediment-laden flash floods occupied the whole Afon Fach valley and were stopped only by the Florentino Ameghino reservoir.

Reservoir water levels rose 11 m in a few days, but there was plenty of storage capacity to hold the water back. A turbidity current from the Afon Fach/Río Chico flowed beneath the surface, however, and some days later reached the dam. The turbines started to deliver an ugly brown discharge into the river downstream, something that was not technically possible to stop. Soon the turbidity wave reached *y Dyffryn* and fresh water supplies were interrupted for nearly two weeks in nearby cities, affecting nearly 200,000 people, Residents had to obtain water from tankers, schools closed, and the price of bottled water increased.[60] The turbidity wave was an extraordinary situation, and engineers raced against time to build large sedimentation reservoirs and restore some drinking water supplies.

Flash floods, droughts and turbidity events are sporadic, and memories are fragile in the Patagonian desert. This means learning to live with the unexpected. It's an everyday practice, and one which stood us in good stead as I observed the March 2019 flash flood alongside Hywel, Stephen and Andrés.

Stephen: Geomorphology in *y Paith*

Since starting PhD studies in Australia over 30 years ago, I have wandered up and down numerous desert ephemeral rivers, streams, creeks, arroyos, washes and wadis worldwide, many of them similar to those draining the Afon Camwy tributary catchments upstream of Ceg y Ffos. The dynamics of many desert rivers – and the wider surrounding desert landscape – are shaped by flash floods that commonly rise quickly with little warning. These flash floods can be beneficial, as they help to maintain highly adapted aquatic ecosystems, recharge aquifers, and supply water, sediment and nutrients to biodiverse floodplains and wetlands. These benefits help support human habitation in the desert, and the environments provide a basis for recreation and geotourism. These benefits, and other ecosystem services of desert rivers, floodplains and wetlands, are widely underappreciated, hence part of the reasons for the workshops in Trelew and Puerto Madryn on our 2019 trip. We need to recognise, inventorise and quantify the different ecosystem services, lest we risk losing more of them through ill-designed and ill-managed grazing, dam building or water abstraction schemes, or through water pollution or urban encroachment. At the same time, flash floods are a significant natural hazard, as they can lead to changes in the size, shape and even location of river channels, and may be associated with infrastructural damage, other economic impacts, and even loss of life. As Gabriel noted in his reflections, in the capricious Patagonian desert, we need to learn to live with the unexpected, but improved information regarding flash flood frequency and size perhaps can make such events a little less unexpected.

Despite extensive experiences with desert rivers, I had always conducted my research during their dry phases, and prior to March 2019, I had not witnessed any actually flowing. I have studied the evidence for flash flood impacts along many different desert rivers; as recorded, for instance, by scour marks on valley walls, gravel and sand deposits, and 'trash lines' of vegetative debris. These different indicators provide vital information on flash flood heights, volumes and velocities; in fact, given the difficulties of maintaining flow gauging stations in deserts, where rainfall and river flows are unpredictable and floods are often damaging when they do occur, these indicators sometimes provide the *only* information sources. The gauging stations maintained by Gabriel and colleagues (Figure 7.2B and 7.2C) are capturing an unusual amount of hydrological and sediment transport data. So as we headed out to the rainy Patagonian desert in Gabriel's truck in March 2019, I was really looking forward to seeing some of the gauging stations in action during small, short-lived floods.

I'll admit that I was disappointed when even the heavy rain initially seemed to do no more than dampen the sandy bed (Figure 7.2B). But as we drove around, it became obvious that on the usually sun-baked surfaces above the rivers, water was starting to pool or be shed downslope (Figure 7.2D), while smaller rivers were starting to flow. We walked across rugged hillsides to larger rivers, many of which were also now starting to flow more forcefully. We were wet through but hey, '*merece la pena*' … it was worthwhile. Finally, I would have my own photos and videos of a water-activated desert to use in future lectures and presentations. But after a few hours, we were all getting a bit cold, and it was time to head back to town for an evening *parrilla* (barbeque).

The desert had other ideas. We returned to cross the previously dry river with its gauging station, but a raging torrent now blocked our exit (Figure 7.2C). With a depth of 1 m and an estimated flow speed of 2 m per second, this was far too dangerous to cross by vehicle, and we had no choice but to wait and hope that the water level would drop sufficiently in a few hours. We made good use of the delayed departure, visiting other flooded rivers to capture more photographs. But as the day wore on, the rain continued in bursts, and as the water level did not seem to be receding very fast, concern set in. Serious thoughts were being given to spending a night in a nearby house that formed part of a small kaolin plant[61] set in the desert, but I didn't dare ask about food provisions. I was more concerned about the almost unthinkable possibilities of not being able to exit the desert even the next day and then missing our flight back to Buenos Aires.

Luckily, just when all seemed lost, at dusk the kaolin plant's owner took us to the river to assess the situation. It was still flowing, albeit slightly less deeply and swiftly than an hour or two beforehand. The plant owner was convinced that it was safe to cross, and without much warning he did just that, barrelling through the turbid current and soft sandy bed to the drier, firmer gravel on the far side. With some trepidation, we followed and also crossed safely. A slow journey then ensued along the rest of the dirt road, with its significant pools of water. But after dark, we finally reached the safety of the tarred road that runs along the more central and northern part of *y Dyffryn*, and we breathed a collective sigh of relief. The desert had given so much to us – a rare geomorphological experience for me personally – but it had very nearly also taken. Eventually, we made it back to Trelew and the *parrilla*, tired but with a tale that will live long.

Final Reflections: Combining Historical and Contemporary Perspectives

What can be gained from combining the archival, the literary and other creative approaches with our personal perspectives, as we have done in this chapter? We can begin to see how the desert has been portrayed historically and how our own desert stories are conditioned both by these portrayals and our own past experiences. The poetry of R. Bryn Williams and Irma Hughes de Jones both emphasise a depiction of the 'civilising' adaptation to the unfamiliar Patagonian desert landscape as heroic and requiring sacrifice, and both refer to the 'miraculous nature' of the work, reflecting a common European settler-colonial outlook. These poems are important records that influence how Welsh speakers or those with Welsh connections, both in Argentine Patagonia and elsewhere, remember the hardships endured by the first colonists, the breakthrough that irrigation represented, and the Florentino Ameghino dam-building. Williams's ode in particular, having an extensive reach among Welsh speakers, was published just as interest in Welsh Patagonia began to be revived around *y Wladfa's* centenary in 1965.[62] When considered alongside the poet's status and wider prose work on Patagonia, the ode likely had a significant impact not only on the perception of *y paith* specifically but also on deserts generally.

We can also appreciate anew the importance of water in desert landscapes, not only water in deficit (as expressed by some of the early colonists) but also water (and sometimes sediment) in excess. The duality of water in desert landscapes is very evident. The rivers provided exploration routes, first from the New Bay to *y Dyffryn* as some colonists sailed south and then came up the mouth of the Afon Camwy to the valley, and later from *y Dyffryn* to the Andes along valleys such as the Afon Camwy and Afon Fach during John Daniel Evans's expeditions. But water in excess would both enforce mobility, as people would have to move away from flooded areas to safety, and also hinder mobility by blocking travel routes. Even during Evans's expeditions, rivers were obstructions as well as creators of paths. Our experience in March 2019 underlined this very directly: the riverbeds very quickly transformed from being routes that we could drive across and walk along, to barriers to movement that almost had left us stranded on *y paith*.

In these archival, literary and creative pieces, and in our personal perspectives, the storied Patagonian desert could certainly be viewed in the context of the Anthropocene.[63] The construction of irrigation canals and a major dam – two significant anthropogenic alterations to a desert region's hydrology – have enabled a measure of adaptation but also led to longer term environmentally damaging consequences. Irrigation has enabled arable agriculture in the desert, but the concomitant raising of groundwater in places has resulted in soil salinisation, which even over half a century ago had started to lower crop yields.[64] Reductions in peak river flows downstream of the dam have provided a heightened sense of security for residents of *y Dyffryn* and led to urban floodplain encroachment, but the associated contraction in the size of the Afon Camwy's channel ironically has increased flood risk, either from dam failure or flash floods originating in smaller catchments downstream. Outside *y Dyffryn*, many desert wetlands – the *mallines* – that provide so many important ecosystem services on *y paith*, are suffering from overgrazing or overabstraction of water. All these negative trends are direct consequences of the 'success' of the apparent adaptation to the desert environment, which is so prevalent in the literary depictions of *y Wladfa*, and which have been judged worthy of remembering through that literature. It could be argued that, in a very similar way to the Elan Valley landscape of transformed land cover and large dams,[65] *y Dyffryn* can be viewed as an 'anthroposcenic landscape' in two senses of the term. Matless refers to an anthroposcenic landscape as one that is 'emblematic' of processes currently at work in our transformed environment.[66] Griffiths et al. use the term with a different meaning, namely a heavily modified landscape, but one that might be positively valued by society.[67] The irrigation canals and associated agricultural landscapes in *y Dyffryn*, for instance, feature widely in popular culture, such as in Kyffin Williams's paintings[68] or in numerous photographic essays[69] and film,[70] and also form the basis for tourism ventures. Similarly, the Florentino Ameghino dam and impounded reservoir is a popular day trip for locals living in *y Dyffryn*[71] and also forms a short, welcome diversion for many tourists travelling westwards across *y paith* to the Andes.

Furthermore, alongside the desert's environmental transformation as recorded in the archival material and in literature, the social transformation – the impact on the Indigenous communities of *y paith* – is also very apparent. Although the depiction of the Indigenous peoples in Williams's ode has some sense of the paternalistic attitude of the Welsh colonists, it does recognise some of the wrongs done to those communities, reflecting the complexity of the Welsh colonists' desert story. Although this is not explored in the poetry, the ironic contrast between the typical historical European settler-colonial view of having to wrest control over the desert environment, and the Indigenous ways of living in and with the landscape in a less negatively impactful way, is striking. Consequently, and going beyond the Patagonian context, these previous contributions and our own findings contribute to a broader, more general understanding of the legacies of colonialism, which are now being re-evaluated in the context of the rapid environmental, social and economic changes strongly associated with the Anthropocene.

Perhaps our combination of these different approaches to the story of *y paith* is like watching those rivulets and streams of water that we saw cascading off the Patagonian hillsides in March 2019 (Figure 7.2D): rivulets and streams that meander and coalesce to create a new, stronger, louder flow. That flow is turbulent and non-linear and erodes the soil. The flow carries individual grains of sediment and gradually creates new channels and new landforms, just as spending time in the Patagonian desert provides individual experiences that shape perceptions and, together, create new stories.

Acknowledgements

The British Academy/Leverhulme Trust and the Sir Ernest Cassel Educational Trust supported Hywel and Stephen's 2014 fieldwork through a Small Research Grant (SG122351), and the British Council supported our 2019 workshop through a Higher Education Links Grant. AU and UNPSJB have supported ongoing activities through a Memorandum of Understanding. Ceris Gruffudd assisted with developing Patagonian links, and Teithiau Tango organised travel. The people of y Wladfa provided warm welcomes, hospitality, and assistance. Gareth Edwin assisted with figures, and the editors and anonymous reviewer provided comments that developed the text. R. Bryn Williams's poem first appeared in Cyfansoddiadau a Beirniadaethau Eisteddfod Genedlaethol Frenhinol Cymru, Abertawe a'r cylch, 1964, published by Llys yr Eisteddfod Genedlaethol, and Irma Hughes de Jones' poems appeared in Edau Gyfrodedd, published by Gwasg Gee. Both publishers generously allowed us to reproduce excerpts and Hywel's translations.

Notes

1 University of Wales Dictionary. 2022. "Paith". University of Wales Dictionary, accessed February 1, 2022. https://geiriadur.ac.uk/gpc/gpc.html.
2 A river catchment is an area of land, bounded by a watershed of higher topography, from which water drains into a river.
3 Joseph Downes, *The Mountain Decameron, Volume 3* (Richard Bentley. 1836).
4 Ronald Paul. "'A Prophet Out of Gwalia:' The Poetry of Idris Davies and Postcolonial Wales." *Zeitschrift für Anglistik und Amerikanistik*, 57, no. 4 (2009): 375–387. https://doi.org/10.1515/zaa.2009.57.4.375
5 Michael Welland, *The Desert: Lands of Lost Borders* (Reaktion Books, 2015).
6 Emrys Bowen, "The Welsh Colony in Patagonia 1865–1885: A Study in Historical Geography." *The Geographical Journal*, 132, no. 1 (1966): 16–27. https://doi.org/10.2307/1793048; David H. Rhys, "*A Geographic Study of the Welsh Colonization in Chubut, Patagonia.*" (PhD diss., University of California, 1976).
7 Glyn Williams, *The Desert and the Dream: A Study of Welsh Colonization in Chubut, 1865–1915* (Cardiff: University of Wales Press, 1975).
8 Bowen, "The Welsh Colony in Patagonia".
9 Williams, *The Desert and the Dream.*
10 Geraldine Lublin, "Y Wladfa: gwladychu heb drefedigaethu?" *Gwerddon* 4 (2009): 8–23; Lucy Taylor, "Global Perspectives on Welsh Patagonia: The Complexities of Being Both Colonizer and Colonized." *Journal of Global History* 13, no. 3 (2018): 446–468. https://doi.org/10.1017/S1740022818000232
11 Taylor, "Global Perspectives on Welsh Patagonia", 447.
12 Rhys, "*A Geographic Study of the Welsh Colonization in Chubut, Patagonia*".
13 Lublin, "Y Wladfa"; Geraldine Lublin, *Memoir and Identity in Welsh Patagonia: Voices from a Settler Community in Argentina* (University of Wales Press, 2017); Lucy Taylor, "Welsh–Indigenous Relationships in Nineteenth Century Patagonia: 'Friendship' and the Coloniality of Power." *Journal of Latin American Studies*, 49, no. 1 (2017): 143–168; Taylor, "Global Perspectives on Welsh Patagonia"; Lucy Taylor, "The Welsh Way of Colonisation in Patagonia: The International Politics of Moral Superiority." *The Journal of Imperial and Commonwealth History*, 47 no. 6 (2019): 1073–1099.https://doi.org/10.1080/03086534.2019.1576836
14 Taylor, "Welsh–Indigenous Relationships".
15 Geraldine Lublin, "Adjusting the focus: looking at Patagonia and the wider Argentine state through the lens of settler colonial theory." *Settler Colonial Studies*, 11, no. 3 (2021): 386–409. https://doi.org/10.1080/2201473X.2021.2001961
16 For example, Hywel M. Griffiths, and Eurig Salisbury, "'The Tears I Shed Were Noah's Flood': Medieval Genre, Floods and the Fluvial Landscape in the Poetry of Guto'r Glyn." *Journal of Historical Geography* 40 (2013): 94–104. https://doi.org/10.1016/j.jhg.2012.11.008; Hywel M. Griffiths and Stephen Tooth, "Remembering and Forgetting Floods and Droughts: Lessons from the Welsh Colony in Patagonia." *Cultural Geographies*, 28, no. 2 (2021): 341–361. https://doi.org/10.1177/1474474020963135; Hywel

M. Griffiths, Eurig Salisbury and Stephen Tooth "'May God Place a Bridge Over the River Tywi': Interrogating Flood Perceptions and Memories in Welsh Medieval Poetry." In *Cultural Histories, Memories and Extreme Weather*, eds. Georgina Enfield and Lucy Veale (Routledge, 2017a), 93–111. https://doi.org/10.4324/9781315461458

17 Clare Madge, "On the Creative (Re)Turn to Geography: Poetry, Politics and Passion." *Area*, 46, no. 2 (2014): 178–185. https://doi.org/10.1111/area.12097; Gabriel Eshun and Clare Madge, "Poetic World-Writing in a Pluriversal World: A Provocation to the Creative (Re) Turn in Geography." *Social and Cultural Geography*, 17, no. 6 (2016): 778–785. https://doi.org/10.1080/14649365.2016.1156147

18 Norah Barlow (Ed.), *Charles Darwin's Diary of the Voyage of H.M.S. Beagle* (Cambridge University Press, 1933), 159–233.

19 Hugh Hughes, *Llawlyfr y Wladychfa Gymreig* (L. Jones & Co., 1862).

20 Rhys, "*A Geographic Study of the Welsh Colonization in Chubut, Patagonia.*"

21 Rhys, "*A Geographic Study of the Welsh Colonization in Chubut, Patagonia.*"

22 Mari Emlyn, *Llythyrau'r Wladfa, 1865–1945* (Gwasg Carreg Gwalch, 2009), 37; translated by Hywel from the original Welsh.

23 Welland, *The Desert: Lands of Lost Borders*.

24 Williams, *The Desert and the Dream*.

25 Griffiths and Tooth, "Remembering and Forgetting Floods and Droughts"; Gabriel Kaless, "Crónica de las inundaciones en el valle y la meseta del chubut. Desde los relatos de los pioneros galeses a los informes técnicos." *Parrafos Geograficos*, 20, no. 2 (2021): 51–104; Gabriel Kaless, Héctor Malnero, Oscar Andres Frumeto, and Miguel Pascual. "When It Rains in the Desert of Patagonia: Heavy Rains, Sediment Laden Flash Floods and Hazard Management Challenges." *Authorea*, October 27, 2021. https://doi.org/10.22541/au.163534875.55080666/v1

26 Fernando Williams, *"Entre el desierto y el jardín." Viaje, literatura y paisaje en la Colonia Galesa de la Patagonia* (Prometeo, 2010).

27 Domingo F. Sarmiento, *Facundo; or Civilization and Barbarism: Life in the Argentine Republic in the Days of the Tyrants* (Hafner Publishing Co., 1845).

28 Williams, Fernando. *"Entre el desierto y el jardín."*

29 Lublin, "Y Wladfa".

30 Wallace Stegner, *Beyond the Hundredth Meridian: John Wesley Powell and the Second Opening of the West* (Penguin Books, 1954); Stephen J. Pyne, *How the Canyon Became Grand: A Short History* (Viking, 1998).

31 Williams, Fernando. *"Entre el desierto y el jardín."*

32 Paul W. Burt (Ed.), *Bywyd a Gwaith John Daniel Evans El Baqueano* (Gwasg Carreg Gwalch, 2004).

33 Burt, *Bywyd a Gwaith John Daniel Evans*, 193.

34 e.g. Burt, *Bywyd a Gwaith John Daniel Evans*, 135.

35 Welland, *The Desert: Lands of Lost Borders*.

36 Hywel M. Griffiths, "Placing y Paith in Poetry." *Poetry Wales*, 51, no. 1 (2015): 26–28.

37 Griffiths and Tooth, "Remembering and Forgetting Floods and Droughts."

38 Richard B. Williams, "Patagonia". In *Cyfansoddiadau a Beirniadaethau Eisteddfod Genedlaethol Frenhinol Cymru*, ed. Lewis Evans (Llys yr Eisteddfod Genedlaethol: 1964), 19–26.

39 Walter A. Brooks, "The Welsh Language in Patagonia: A Brief History." *British Council*, October 25, 2018, accessed October 28, 2023. https://www.britishcouncil.org/voices-magazine/welsh-language-in-patagonia-and-wales.

40 See Mererid Hopwood, *Singing in Chains: Listening to Welsh Verse* (Gwasg Gomer, 2004).

41 Williams, "Patagonia", 19.

42 Williams, "Patagonia", 19–20.

43 Williams, "Patagonia", 20.

44 Williams, "Patagonia", 20.

45 Williams, "Patagonia", 21.

46 Williams, "Patagonia", 22.
47 Williams, "Patagonia", 23.
48 Williams, "Patagonia", 23.
49 Williams, "Patagonia", 23.
50 Williams, "Patagonia", 23.
51 Williams, "Patagonia", 23.
52 Williams, "Patagonia", 24.
53 Williams, "Patagonia", 25.
54 Williams, "Patagonia", 25.
55 Rhiannon Marks, "Irma Ariannin, Cymru a'r 'wlad lle cyferfydd cyfandiroedd'." *Ysgrifau Beirniadol*, 34 (2016): 13–43.
56 Irma Hughes de Jones, *Edau gyfrodedd: detholiad o waith Irma Hughes de Jones*, ed. Cathrin Williams (Gwasg Gee, 1989), 54.
57 Hughes de Jones, *Edau Gyfrodedd*, 43–47.
58 Sarmiento, *Facundo; or Civilization and Barbarism.*
59 Griffiths and Tooth, "Remembering and Forgetting Floods and Droughts"; Kaless, "Crónica de las inundaciones."
60 James McCarthy, "Welsh Patagonia Is Slowly Recovering after Being Left Devastated by Flooding." *WalesOnline*, June 10, 2019, accessed October 28, 2023. https://www.walesonline.co.uk/news/wales-news/welsh-patagonia-slowly-recovering-after-13158044
61 Kaolin is one of many minerals mined in Patagonia. Plans for large, so-called 'mega-mines' in the Patagonian region have been the subject of environmental protests.
62 Brooks, "The Welsh Language in Patagonia".
63 Aidan Tynan, *The Desert in Modern Literature and Philosophy: Wasteland Aesthetics* (Edinburgh University Press, 2020), https://doi.org/10.1080/14688417.2021.1894745
64 John F. Bergmann, "Soil Salinization and Welsh Settlement in Chubut, Argentina." *Cahiers de géographie du Québec*, 15, no. 35 (1971): 361–369. https://doi.org/10.7202/020966ar
65 Hywel M. Griffiths, Gavin Goodwin, Tyler Keevil, Eurig Salisbury, Stephen Tooth and Dewi Roberts. "Searching for an Anthropo(s)cene in the Uplands of Mid Wales." *GeoHumanities*, 3, no. 2 (2017b): 567–579. https://doi.org/10.1080/2373566X.2017.1329631
66 David Matless, "The Anthroposcenic." *Transactions of the Institute of British Geographers*, 42, no. 3 (2017): 363–376. https://doi.org/10.1111/tran.12173. p. 363.
67 Griffiths et al., "Searching for an Anthropo(s)cene".
68 Paul Joyner (Ed.), *Gwladfa Kyffin: Kyffin in Patagonia* (The National Library of Wales, 2004).
69 e.g. Ed Gold, *Patagonia: Byd Arall/Otro Mundo/Another World* (Gwasg Gomer, 2012); Eirionedd A. Baskerville, *Patagonia 150: Yma I Aros/Here to Stay/Aquí Para Quedarse* (y Lolfa, 2015).
70 Jennifer Wood, "Signifying the Nation: (In) Communication, Absence and National (Be) Longing in Marc Evans's Patagonia." In *Spaces of Longing and Belonging: Territoriality, Ideology and Creative Identity in Literature and Film*, eds. Le Juez, Brigitte and Bill Richardson (Brill, 2010), 210–232. https://doi.org/10.1163/9789004402935_014
71 Griffiths and Tooth, "Remembering and forgetting floods and droughts."

Bibliography

Barlow, Norah (Ed.). *Charles Darwin's Diary of the Voyage of H.M.S. Beagle*. Cambridge University Press, 1933.

Baskerville, Eirionedd A. *Patagonia 150: Yma I Aros/Here to Stay/Aquí Para Quedarse*. Y Lolfa, 2015.

Bergmann, John F. "Soil Salinization and Welsh Settlement in Chubut, Argentina." *Cahiers de géographie du Québec* 15, no. 35 (1971): 361–369. https://doi.org/10.7202/020966ar.

Bowen, Emrys. G. "The Welsh Colony in Patagonia 1865–1885: A Study in Historical Geography." *The Geographical Journal* 132, no. 1 (1966): 16–27. https://doi.org/10.2307/1793048.

Brooks, Walter A. "The Welsh Language in Patagonia: A Brief History." *British Council*, October 25, 2018, accessed October 28, 2023. https://www.britishcouncil.org/voices-magazine/welsh-language-in-patagonia-and-wales.

Burt, Paul W. (Ed). *Bywyd a Gwaith John Daniel Evans El Baqueano*. Gwasg Carreg Gwalch, 2004.

Downes, Joseph. *The Mountain Decameron, Volume 3*. Richard Bentley, 1836.

Emlyn, Mari. *Llythyrau'r Wladfa, 1865–1945*. Gwasg Carreg Gwalch, 2009.

Eshun, Gabriel, and Clare Madge. "Poetic World-Writing in a Pluriversal World: A Provocation to the Creative (Re) Turn in Geography." *Social and Cultural Geography* 17 no. 6 (2016): 778–785. https://doi.org/10.1080/14649365.2016.1156147.

Gold (Ed.). *Patagonia: Byd Arall/Otro Mundo/Another World*. Gwasg Gomer, 2012.

Griffiths, Hywel M. "Placing y Paith in Poetry." *Poetry Wales* 51, no. 1 (2015): 26–28.

Griffiths, Hywel M., and Eurig Salisbury. "'The Tears I Shed Were Noah's Flood': Medieval Genre, Floods and the Fluvial Landscape in the Poetry of Guto'r Glyn." *Journal of Historical Geography* 40 (2013): 94–104. https://doi.org/10.1016/j.jhg.2012.11.008.

Griffiths, Hywel M., and Stephen Tooth. "Remembering and Forgetting Floods and Droughts: Lessons from the Welsh Colony in Patagonia." *Cultural Geographies* 28, no 2 (2021): 341–361. https://doi.org/10.1177/1474474020963135.

Griffiths, Hywel M., Eurig Salisbury, and Stephen Tooth. "'May God Place a Bridge over the River Tywi': Interrogating Flood Perceptions and Memories in Welsh Medieval Poetry." In *Cultural Histories, Memories and Extreme Weather*, eds. Georgina Enfield and Lucy Veale. Routledge, (2017a), 93–111. https://doi.org/10.4324/9781315461458.

Griffiths, Hywel M., Gavin Goodwin, Tyler Keevil, Eurig Salisbury, Stephen Tooth, and Dewi Roberts. "Searching for an Anthropo(s)cene in the Uplands of Mid Wales." *GeoHumanities* 3, no. 2 (2017b): 567–579. https://doi.org/10.1080/2373566X.2017.1329631.

Hopwood, Mererid. *Singing in Chains: Listening to Welsh Verse*. Llandysul: Gwasg Gomer, 2004.

Hughes, Hugh. *Llawlyfr y Wladychfa Gymreig*. L. Jones & Co., 1862.

Hughes de Jones, Irma. *Edau gyfrodedd: detholiad o waith Irma Hughes de Jones*, ed. Cathrin Williams. Denbigh: Gwasg Gee, 1989.

Joyner, Paul (Ed.). *Gwladfa Kyffin: Kyffin in Patagonia*. The National Library of Wales, 2004.

Kaless, Gabriel. "Crónica de las inundaciones en el valle y la meseta del chubut. Desde los relatos de los pioneros galeses a los informes técnicos." *Parrafos Geograficos* 20, no. 2 (2021): 51–104.

Kaless, Gabriel, Oscar Andres Frumento, Érico Hernán Bianchi, Maria Jesus Chachero, Rodrigo Bastida, and Sebastian Iglesias. "Cuando en el desierto patagónico los arroyos braman. Evento extraordinario de abril de 2017 en la cuenca del Arroyo Perdido." *Capitulo* 21 (2019): 273–288.

Kaless, Gabriel, Héctor Malnero, Oscar Andres Frumeto, and Miguel Pascual. "When It Rains in the Desert of Patagonia: Heavy Rains, Sediment Laden Flash Floods and Hazard Management Challenges." *Authorea*, October 27, 2021. https://doi.org/10.22541/au.163534875.55080666/v1

Lublin, Geraldine. "Y Wladfa: gwladychu heb drefedigaethu?" *Gwerddon* 4 (2009): 8–23.

Lublin, Geraldine. *Memoir and Identity in Welsh Patagonia: Voices from a Settler Community in Argentina*. University of Wales Press, 2017.

Lublin, Geraldine. "Adjusting the Focus: Looking at Patagonia and the Wider Argentine State Through the Lens of Settler Colonial Theory." *Settler Colonial Studies* 11, no. 3 (2021): 386–409. https://doi.org/10.1080/2201473X.2021.2001961.

Madge, Clare. "On the Creative Re(Turn) to Geography: Poetry, Politics and Passion." *Area* 46, no. 2 (2014): 178–185. https://doi.org/10.1111/area.12097.

Marks, Rhiannon. "Irma Ariannin, Cymru a'r 'wlad lle cyferfydd cyfandiroedd'." *Ysgrifau Beirniadol* 34 (2016): 13–43.

Matless, David. "The Anthroposcenic." *Transactions of the Institute of British Geographers* 42, no. 3 (2017): 363–376. https://doi.org/10.1111/tran.12173.

McCarthy, James. "Welsh Patagonia Is Slowly Recovering after Being Left Devastated by Flooding." *Wales Online*, June 10, 2019, accessed October 28, 2023. https://www.walesonline.co.uk/news/wales-news/welsh-patagonia-slowly-recovering-after-13158044

Paul, Ronald. "'A Prophet Out of Gwalia': The Poetry of Idris Davies and Postcolonial Wales." *Zeitschrift für Anglistik und Amerikanistik* 57, no. 4 (2009): 375–387. https://doi.org/10.1515/zaa.2009.57.4.375.

Pyne, Stephen J. *How the Canyon Became Grand: A Short History*. Viking, 1998.

Rhys, David H. *A Geographic Study of the Welsh Colonization in Chubut, Patagonia* (PhD diss.). University of California, 1976.

Sarmiento, Domingo F. *Facundo; or Civilization and Barbarism Life in the Argentine Republic in the Days of the Tyrants*. Hafner Publishing Co., 1845.

Stegner, Wallace. *Beyond the Hundredth Meridian: John Wesley Powell and the Second Opening of the West*. Penguin Books, 1954.

Taylor, Lucy. "Welsh–Indigenous Relationships in Nineteenth Century Patagonia: 'Friendship' and the Coloniality of Power." *Journal of Latin American Studies* 49, no. 1 (2017): 143–168.

Taylor, Lucy. "Global Perspectives on Welsh Patagonia: The Complexities of Being Both Colonizer and Colonized." *Journal of Global History* 13, no. 3 (2018): 446–468. https://doi.org/10.1017/S1740022818000232.

Taylor, Lucy. "The Welsh Way of Colonisation in Patagonia: The International Politics of Moral Superiority." *The Journal of Imperial and Commonwealth History* 47, no. 6 (2019): 1073–1099. https://doi.org/10.1080/03086534.2019.1576836.

Tynan, Aidan. *The Desert in Modern Literature and Philosophy: Wasteland Aesthetics*. Edinburgh University Press, 2020. https://doi.org/10.1080/14688417.2021.1894745.

University of Wales Dictionary. "Paith." University of Wales Dictionary, 2022, accessed February 1 2022. https://geiriadur.ac.uk/gpc/gpc.html.

Welland, Michael. *The Desert: Lands of Lost Borders*. Reaktion Books, 2015.

Williams, Fernando. '*Entre el desierto y el jardín.' Viaje, literatura y paisaje en la Colonia Galesa de la Patagonia*. Prometeo, 2010.

Williams, Glyn. *The Desert and the Dream: A Study of Welsh Colonization in Chubut, 1865–1915*. University of Wales Press, 1975.

Williams, Richard B. "Patagonia." In *Cyfansoddiadau a Beirniadaethau Eisteddfod Genedlaethol Frenhinol Cymru*, ed. Evans, Lewis. Llys yr Eisteddfod Genedlaethol, 1964, 19–26.

Wood, Jennifer. "Signifying the Nation: (In) Communication, Absence and National (Be) Longing in Marc Evans's Patagonia." In *Spaces of Longing and Belonging: Territoriality, Ideology and Creative Identity in Literature and Film*, ed. Le Juez, Brigitte and Bill Richardson. Brill, 2010, 210–232. https://doi.org/10.1163/9789004402935_014.

8

PARCHED POETICS

Disarticulations from a Desert

Gretchen Ernster Henderson

Parched Poetics

Listen. Smell the air. Nudge your fingers and toes in sand and sediment. Feel movements across your body: muscle tensions, posture, temperatures of heat, or cool. Imagine phenomena circulating with other bodies, not only human and animal but also bodies of land and bodies of water. Consider lands of little water, so-called deserts.

In the high desert of Utah, at the edge of the arid Great Basin, a body of water called Great Salt Lake has been likened to a heartbeat, oxygenated by arterial rivers and water cycles. In recent years, the lake has receded from drought. As islands rise, the lake sinks in a basin of mountainous ranges. Natural inscriptions mark the landscape: with etched shorelines of ancient lake levels, crisscrossed by quaking faults, beside natural tar seeps that smudge the edge of this reputedly dead sea.

Tar seeps are pools of raw oil, nicknamed "death traps," seeping up from tectonic fractures and spreading across the earth's surface like tacky flypaper. An unsuspecting bird or animal wandering through a melted seep can get fatally stuck.

The tar seeps in the remote north arm of Great Salt Lake lie near Robert Smithson's earthwork of *Spiral Jetty*. Bulldozed into shape in 1970, the artwork unfurls a path of salt-encrusted, black basalt—15 feet wide, and 1,500 feet long—counterclockwise three times into the seeming sea. *Spiral Jetty* has been considered surreally beautiful and also an ugly scar. Smithson selected this site deliberately for its proximity to the tar seeps, littered with scars of abandoned attempts at oil drilling. In a fairly wild landscape, the receding lake has left both the spiral and seeps relatively beached.

Around *Spiral Jetty*, around the tar seeps, the living lake moves in seeming stillness. Microbial life swims, finless, in colorful saline saturation, immune to the

DOI: 10.4324/9781003397212-11

burning sun. Microseisms emanate as faint tremors from the slowly moving planet. Storms whip up waves, and in winter, winds can chill to the bone. Even on a seemingly still, sunny day, there is movement: buoyant in dense saltwater, evaporating into the atmosphere. The air can stink from bacteria, natural asphalt, and decaying matter. Vapor condenses, precipitating through cycles across ranges back to the basin, through snowmelt and river runoff, where birds migrate to marshes. The air can pulse with the flight of pelicans—a huffing *wingbeat-wingbeat-wingbeat*—that hushes you in your tracks. The site lies near the convergence of two of the major migratory bird flyways of North America.

Looking outward, *Spiral Jetty* offers a focal point on this evolving climatic movement. The artwork has been decomposing over a half century. For three decades after its creation, this massive coil of rocks lay underwater. Arising from the receding lake in 2002, *Spiral Jetty* grew from an aura of absence, echoing a mythic whirlpool while becoming a bellwether of drought. Against the lake, the artwork is relatively small. Photographs exaggerate its size and scale, centering what otherwise decentralizes. In aerial photography, *Spiral Jetty* becomes the focal point. A walk on the artwork unfocuses its shape as the view turns outward: to the lake, to the mountains, to clouds and sky.

Although I originally came to Rozel Point because of *Spiral Jetty* in 2017, since then I have accompanied scientists on their studies of the adjacent tar seeps and surrounding ecosystem. I have grown to perceive both the art spiral and tar seeps as earthworks—one man-made, the other nature-made—evolving in dialogue. Art and tar, salt and water, feathered birds, and weathered bones have congealed around my perceptions of life and death, injury and healing, degeneration and regeneration. Articulation and disarticulation. Deserts are associated with dying, but the lake's natural agency asserts relationally with the tar seeps, microbial lives, birds, and other species that shift my own somatic experience in kind—to perceive this "dead sea" as deeply alive and a watershed for thinking about any overlooked place.

As agricultural runoff, toxic dumps, pollution, extractions, and other environmental threats impede Great Salt Lake's life, my perceptions spiral farther afield. Birds who feed at the lake migrate and take toxins elsewhere. Down from the dirt parking lot by *Spiral Jetty*, past the tar seeps and rotting pilings, a rutted road links the lake through fenced ranchlands to the Golden Spike Historical Site at Promontory, where the first cross-country railroad connected. Widening frames of reference spiral across the basin and ranges through Indigenous archaeological and modernized human settlements,[1] natural resource extractions, military-industrial installations, interplanetary simulators, and radio observatories, which tune to listen to the mini-spiral of gas at the center of our spiraling galaxy. The lake's contingency interrelates with other places and stakes, from water cycles and droughts of the American West to astrobiological analogues on Mars.

The tensions at Rozel Point hold my attention and raise questions about aesthetics and languages of arid environments, wastelands, and other overlooked places across a desertifying planet. Everything is interconnected. The classifications that we use to describe any environment are always partial—in the sense of both incomplete and inclined—reflecting our histories and fragmented attempts to communicate and connect, with biases and miscommunications, missed or disconnections. Disciplinary languages can illuminate but also obfuscate. Just as bones can be articulated, language also can be disarticulated and fossilize. Etymologies signal origin stories from places, eliding with displacements and diasporas. Meanings of words dry up, as if parched, shadowing former vibrance. Fieldwork grows from a field, not only a physical place but also an evolving practice. Gaps invite emergences from the evolving world.

For over a half-decade now, I have visited *Spiral Jetty* repeatedly over different seasons and reframed this parched place through varied media (photography, video, GoPro, drone, tar prints, writing, more). Yet attempts to represent or border land art—across photographs, film, writing, archives—reveal the limitations of any medium or genre, expose the view of a creator, and lose what lies at the edge or outside the frame. To some degree land art's resistance to being reproduced—even literally, through copyright restrictions that limit use of others' photographs as well as our own—forces attention beyond the object to the place it occupies, what lies adjacent and overlooked, including natural tar seeps and the lake itself, and gaps in our perceptions.

Considering a poetics of space, how does language expand or hinder a sense of place? How does a human imagination stay primed to renew ways of perceiving, to pay attention and care? How can human bodies as sensory registers better attune with other bodies: animal, botanical, and beyond?

In the context of this anthology of *Storied Deserts: Reimagining Global Arid Lands*, my poetics of aridity—what I offer as a "parched poetics"—entwine with other deserts across the world. Arid environments have often borne associations of "dead," "barren," "withered," "ugly," "lacking interest or meaning" to describe living places. As meanings erode through accumulation, my offering of "seeping through the tar" is a de/recomposing essay akin to an arid poetica, reflecting a place through the limits of classification. (That includes a GPS coordinate, which geolocates a sense of place through disembodied data.) An essay assays akin to mining and extraction. Since deserts are often framed as spaces of negation, this poem-that-is-not is a gesture to declassify. Vocabularies can be viscous.

As a tar seep melts and freezes, my experiences at Great Salt Lake compress a story of dying to reflect how much is living. My longer term engagement with this high desert melts from a larger project, *Life in the Tar Seeps: A Spiraling Ecology from a Dying Sea* (Trinity University Press 2023).[2]

It may be too easy and almost cliché to associate a poetics of aridity with a drought of language, a parched poetics, eroding to spareness—like an artwork cast in relief on the edge of a retreating shore at Great Salt Lake. Disarticulating, words turn material and fossilize. But fossils can have afterlives, and speculative futures grow from excavated histories. As we live in the present, languages enliven through listening. In reimagining, arid environments assert a natural agency that is part of a larger story that may be written and rewritten yet cannot be written off. From sands to cells, dust to stardust, humans live for a precious blip in the grand span of this planet in a spiraling universe. It can be easy to neglect our imprint. As we constantly renegotiate our relationship with the living Earth, a tar seep always reminds us to watch where we step. Where do you get stuck?

Seeping through Tar: The Limits of Classification

41°, 25′, 50″ N; 112°, 39′, 41″ W
(*From a tar seep at Rozel Point, Great Salt Lake, Utah, North America, Earth*)

FIGURE 8.1 "Field Note: Aerial view of tar seeps, abandoned oil jetty and mudflats," Rozel Point, Great Salt Lake, Utah, 06/12/21.

Drone photograph by Gretchen Ernster Henderson.

: abandoned, accident, aerial, aesthetics, agency, agricultural runoff, alive, American West, anger, Anthropocene, archaeology, archive, arid, arterial, artwork, apocalyptic, articulation, asphalt, *Asphalt Rundown*, **In the high desert of Utah at Great Salt Lake, a series of tar seeps spreads across the earth.** bacteria, basalt, basin, Bear River Massacre, Bear River Migratory Bird Refuge, Binagadi asphalt lake (Azerbaijan), biodiversity, biorhythm, birds, bitumen, black gold, bones, brea, breathing, brine shrimp, burial, burning water, byproduct, **Nicknamed 'death traps,' also called *oil* or *petroleum seeps*, tar seeps act like sticky flypaper and creep up from tectonic fractures at seismic spots across the planet.** camera trap, capped well, carbon dioxide, Carpinteria Tar Pits and Oil Seeps (CA), cattle guard, caulk, causeway, caves, CLUI (Center for Land Use Interpretation), classification, climate change, climate crisis, circumnavigate, collective, colonial-settlement, colonization, Colorado River Watershed, combustion, composition, conservation, constellation, crude oil, cultural history, **Over centuries, tar has accumulated fatal and extractive associations, often accompanied by oil drilling.** danger, dark skies, death trap, decenter, decolonize, decomposition, deep time, derrick, desert, desertification, Dia Art Foundation, digital age, disarticulate, disaster, DNR–FFSL (Department of Natural Resources–Forestry, Fire and State Lands), documentation, drains, drills, drips, drone, drought, dry playa, dust, dystopia, **Culturally, tar also has stuck things together** Earth, earthwork, ecology, ecotone, elegy, elsewhere, embodied, empathy, energy, entropic, environment, evaporation ponds, evidence, evolution, excavation, extinction, extraterrestrial, **breaking down limited classifications of death** fallout, fault, fault line, feathers, field guide, fill-in-the-blank, fissure, flammable, flow, flyway, focal point, footprints, Fort Sill Tar Pits (OK), fossil, fossilize, fossil fuels, fragment, frozen, futures, **to reveal underlying life, telling a tale of place-in-time:** gas, geology, geothermal, Golden Spike, Goshute, gravity, Great Basin, Great Salt Lake, GSLEP (Great Salt Lake Exploration Platform), GSLI (Great Salt Lake Institute), grief, ground, groundwater, gull, Gunnison Island, gypsum, **where an unsuspecting animal (like a pelican) following a trail over seasons (as tar freezes and melts) can get stuck.** habitat, halophiles, harvest, hawk, hazard, healing, heartbeat, heat, heliography, here, Holt (Nancy), hope, horizon line, human, human activity, humility, hydrocarbon, **From tar pits in Los Angeles (with ancient mammoths, giant sloths, and saber-toothed cats), to tar seeps in the Talara Desert of Peru (with ancient songbirds), natural asphalt can fatally preserve.** I, igneous, incomplete, Indigenous, Industrial Revolution, injury, injustice, inscription, interdependent, international, intervention, inventory, inversion, IPCC (Intergovernmental Panel on Climate Change), islands, **In the reputedly 'dead sea' of Great Salt Lake, tar seeps lie near the convergence of two of the four major migratory bird flyways of North America** jeopardize, jetty, Jordan River, journalism, jumbled, juncture, juxtaposition, **and beside Robert Smithson's famous earthwork coiling into the seeming sea.** Kalahari Desert (Lake Makgadikgadi, Botswana), kin, knowing, knowledges, **Smithson chose the site for *Spiral Jetty* (1970) because of the tar seeps.** La Brea Tar Pits (CA), lake, Lake Bermudez (Venezuela), Lake Bonneville, lake levels, land art, Land Arts of the American West, landscape, larvae, LDS (Church of Latter Day Saints), leased lands, "leave no trace," liquid, lives, local, lubricant, **Like a massive ink blot on the planet, tar seeps seem to defy language,** manufacture, Mars, McKittrick Tar Pits (CA), meander line, metadata, metamorphic, metaphor, methane, microbialite, microenvironments, microseisms, migratory birds, military testing, Milky Way, mine, mineral, mineral pitch, mineral rights, mining, mortal, Mother Earth, muck, mud, mudflat, mummify, myth, **as we grope to articulate qualities:** national, narrative, NASA (National Aeronautics and Space Administration), natural gas, natural resources, nature, negation, Nevada Test Site, non-invasive, non-site, North America, nuclear fallout, **falling back on the fossilized vocabulary of paleontology, to migratory biodiversity of ornithology, to the economic value of the oil.** object, observations, ode, odor, oil, oil company, oil field, oil spill, oil well, ooze, ornithology, osprey, overlooked, overwritten, owl, oxygen, **As I acquire terms**, paleontology, past, pastoral, pathway, paved, pelicans, perception, perspective, petroleum, photography, pipe, pit, pitch, Pitch Lake (Trinidad and Tobago), playa, Pleistocene, poetics of space, point of view, politics, porous, posthumous, precipitation, prehistory, preservative, preserve, primal, project, projection, protect, **the terrain evades classification. A tar seep swallows all languages that attempt to describe it.** quakes, questions, QR-code (Quick Response), **Bacteria break it down.** reclamation, recover, Red List, Refuge, refuge, regional, rejuvenate, relational, reminder, renewable, resilience, resurgence, resurrection, retreat, re-view, river runoff, rock, Rozel Point, **As some bacteria even eat tar, they suggest hope for cleaning up future oil spills, and something more philosophical:** sacred, sacrifice, sagebrush, saline, salt, salt flat, sand, science fiction, sediment sample, "sedimentation of the

FIGURE 8.2 "Field Note: circling star-shaped seep," Rozel Point, Great Salt Lake, Utah, 02/12/2018.

Photograph by Gretchen Ernster Henderson.

mind," seepage, seiche, seismic, shadow, shoreline, Shoshone, sink, slow reading, smell, Smithson (Robert), snowmelt, sodium chloride, solar system, species, speculative, *Spiral Jetty,* stars, steward, sticky, stinky, stone, strata, stuck, subject, subjective, sublime, sulfur, *Sun Tunnels,* surface, surveying, survival, **about the power of microbial species, integral to our life cycles.** Talara Tar Seeps (Peru), Tanque Loma Tar Seep (Ecuador), tar balls, tar pits, tar sands (Canada), tar seeps, tar volcanoes, time, timescale, tracking, travel, travelogue, **Bacteria line our guts, maintain our body chemistry,** UGS (Utah Geological Survey), UMFA (Utah Museum of Fine Arts), US (United States), us, Utah, Ute, **and one day decompose us back to dirt—unless we get stuck in a tar seep. With the receding lake,** valley, value, vanishing, variation, view, viewfinder, viewshed, virtual, viscous, vision, visit, vitality, void, volatile, vulnerable **tar seeps assert interrelation to reveal a 'dead sea' in the desert as deeply alive: a watershed for thinking about any overlooked place.** Wasatch Mountains, water, water diversion, water rights, water runoff, watershed, wave, we, web, Weber River, well, Wendover, West, wetland, wild, wildfire, wildlife, wonder, world, worldview, **Revaluing what has been dismissed as 'dead,' 'ugly,' and derivatives: what other words might shift perceptions of deserts?** xeric, you are here: **and what emerges beyond the limits of language?**

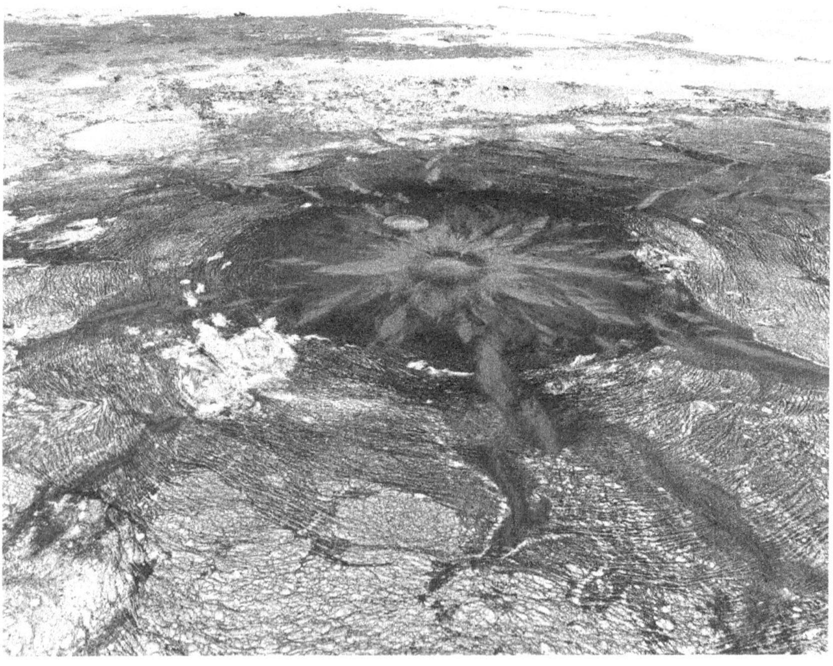

FIGURE 8.3 "Field Note: star-shaped tar seep with barn owls," Rozel Point, Great Salt Lake, Utah, 02/12/2018.

Photograph by Gretchen Ernster Henderson.

Notes

1 With gratitude to longtime Indigenous stewards of the land currently called Utah: I (We) acknowledge that this land, named for the Ute Tribe, is the unceded, traditional, and ancestral homeland of the Shoshone, Paiute, Goshute, and Ute Tribes. I (We) acknowledge the significance of place and the continued existence and contributions of Indigenous people who have lived on and cared for this land for thousands of years. I (We) respect the sovereign relationship between tribes, states, and the federal government, and I (We) affirm the need for more commitment to listening and partnerships with Native Nations and Urban Indian communities. (For more: University of Utah Communications, 'Indigenous Land Acknowledgement' (9 October 2020): https://attheu.utah.edu/facultystaff/indigenous-land-acknowledgement/.) Land Acknowledgements, as words, have limits without actions.

2 To learn more, visit: gretchenhenderson.com/life-in-the-tar-seeps

PART 3

Land

Politics of Belonging

9

COMBATTING DESERTIFICATION AND NARRATING ENVIRONMENTAL CRISIS IN THE UNITED NATIONS

Veronika Kratz

As climate change continues to be pulled into polemic arguments of belief and disbelief, the United Nations (UN) and its bodies, like the Intergovernmental Panel on Climate Change (IPCC), emerge as bastions of reliability. The UN and its associated agencies are especially trusted as scientific authorities on environmental degradation, which they often name as desertification. Ecologist Mark Stafford Smith writes that, despite mounting scientific evidence that desertification as a term is inaccurate, misleading, and often an actively harmful basis for policy-making in drylands, it isn't likely to fall out of popular usage because it remains the UN's main term of reference for land degradation in arid and semi-arid regions.[1] The UN's stamp of authorization on desertification suggests a global consensus that simply doesn't exist.

In order to understand why the UN continues to align desertification with its work on land degradation, drought, and soil health, this chapter investigates how desertification functions as a useful story of environmental crisis. Centering on an exploration of the implicit narratives contained within the UN Conference on Desertification (UNCOD) from 1977 and the UN Convention to Combat Desertification (UNCCD) from 1996, I argue that the UN offers a compelling story of environmental crisis that relies on a simplistic metaphor of desert advance wherein anti-desertification work seeks to halt encroaching desert sands. While this framework of environmental crisis as a conflict between people and land is effective as a communication tool, it limits our ability to understand and respond to land degradation in arid regions by obscuring the colonial history of desertification and perpetuating a misunderstanding of deserts and drylands as bad or broken landscapes. In essence, desertification serves as a popular story of deserts in crisis, but in contrast to the multiple and diverse examples of storied deserts explored elsewhere in this collection, desertification is a singular story that limits our ability to recognize the full complexity of drylands.

DOI: 10.4324/9781003397212-13

In this chapter, I establish the UN's narrative of desertification as a neocolonial concept. Using articles from the *UNESCO Courier* magazine, I demonstrate that core aspects of the colonial narrative are upheld even as authors distance the term from that history using the supposed neutrality of scientific expertise. These ideas inform the 1977 UNCOD, which saw the establishment of desertification as a narrative of environmental crisis, utilizing spectacle and a sense of urgency to communicate the threat to humanity supposedly posed by desert spaces. This narrative is maintained in the UNCCD, which entered into force December 1996. The UNCCD was designed to be more accurate, scientific, and realistic than its predecessor, but despite its appearances of neutrality, it is still based on the metaphor of conflict with a vilified, expanding desert. In cultural products of the UN and its various mitigation efforts, desertification emerges as a broken term of reference for land degradation in arid lands, but a useful example of what is gained and lost in the narration of environmental crisis through spectacular metaphor.

While the policy documents associated with the UNCOD and UNCCD center my discussion, the UN's narrative of desertification is not static, nor is it contained neatly within these texts. Therefore, my analysis also engages with a wider network of archival materials including audio recordings, scientific articles, and personal reflections, all of which demonstrate, replicate, and translate the organization's ideas to a broader public of educators, scientists, non-governmental organizations (NGOs) and public figures. The UNCOD and the UNCCD outline the shape of the UN's narrative of desertification and allow for a clearer analysis of how the topic is treated in this broader UN system, which in turn shapes popular public opinion on desertification itself.

Defining desertification has always been a challenge. In the words of one UN negotiator, the question of the term's definition has "haunted the international community for years,"[2] while a 1985 article from the Food and Agriculture Organization offers this simple definition: "[i]f erosion is the sickness of a land, desertification is its death."[3] There are many more definitions available (though few are so dramatic), but in a general sense, desertification usually refers to land degradation in arid or semi-arid regions resulting from several possible factors, including severe drought and human action. Sometimes this is simplified to the encroachment of desert-like conditions, and sometimes it can refer to a loss of soil productivity in a particular area. The space between these extremes (and a myriad more variations) is wide, and the disparity of the term is part of why it is so troubling: because there is actually a very big difference between a drop in soil productivity and the idea that deserts are encroaching onto farms, pastureland, settlements, and infrastructure. Deserts are not expanding, nor are they representative of degraded land. What is perhaps the most surprising is that these extremes of definition frequently coincide. Even when the term is used to refer to degraded soils in a given area, the word "desertification" invokes the idea of a steady, spreading threat.

The idea of encroaching deserts is tied to the origins of desertification. Geographer Diana K. Davis has written extensively on the emergence of desertification in the French colonies of North Africa, and argues that it should be understood as a *neocolonial* concept that has changed very little since its inception in the eighteenth century.[4] The actual term "desertification" was coined by French colonial forester Louis Lavauden in 1927, but Davis's work illustrates that Lavauden was building on a tradition of European misunderstandings of arid landscapes that saw them as having been degraded by human mismanagement. This narrative functions to not only blame arid conditions on local populations (who have supposedly ruined the land) but also presents deserts themselves as the result of degradation. Ecocritic Tom Lynch has also noted the way that "desertification" positions deserts as conditions "to be avoided," or, when possible, *transformed*, and notes that this is part of the real harm that the metaphorical ideas of desertification can bring to real desert spaces.[5]

The extensive scholarship on the shaky "science" of desertification and its history as a narrative of environmental control enables my own work which investigates desertification from a literary and cultural lens. Accordingly, my work contributes to research that has referred to desertification variously as a narrative, myth, and paradigm. This variation offers the opportunity to make some important observations of its character. For example, geographer Michael Mortimore's work discusses desertification as a paradigm to focus on how this discourse influences disciplinary and institutional approaches to drylands, especially within large organizations like the UN, and national governments.[6] Calling desertification a myth is done to highlight its falseness—specifically, the lack of scientific evidence to support it, as well as the way that it has changed over time to suit the needs of the storyteller.[7] A narrative or story is different again. Stories of desertification position actors, environments, and timescales in particular ways. For example, economist Jeremy Swift describes desertification as a development narrative wherein rural populations are victimized and modernization/development is situated as a solution to their plight.[8] In my examination of desertification through the work of the UN, I focus on the utilization of desertification as a desert story that centers on environmental crisis to create a spectacle out of drought and dryness in order to demand urgency and attention.

Today, desertification is also defined as a global environmental crisis resulting from human action. It is part of our contemporary discourse on climate change, sitting alongside other concerns such as pollution, melting sea ice, biodiversity loss, deforestation, and sea level rise. In this I want to be clear that my goal is not to question the validity of the varied environmental emergencies we face on this planet—far from it. But each of these concerns is made up of cultural narratives that must be understood if we hope to move forward. The cultural narratives of environmental crises determine how we understand them, how we approach them, and generally, how we see humanity in relation to them. What I offer in this chapter is an examination of the narrative of desertification developed and upheld by the UN since the 1970s.

Desertification and Colonial Environmental Narratives in Africa

While the UN is largely responsible for the current use of the term as we know it today, desertification as a concept emerged from colonial governments in Africa where it was utilized as justification for foreign control over those lands and peoples. In order to analyze the narratives of desertification that have been established by the UN since the 1970s, it is crucial to recognize how much of the story has been created by and remains relatively unchanged from this colonial context. Looking at the articles written for the June 1977 edition of the *UNESCO Courier*, a magazine designed to translate UN work for a global public audience, it is clear that the colonial narrative is maintained. Throughout the publication, there is a generally held view that all deserts are degraded spaces in need of transformation, as it blames local people, especially nomads living in dryland environments, for land degradation. Scientists and experts working for UNESCO try to distance the idea of desertification from this history, as the term is increasingly associated with science and development rather than colonial land management, despite the interrelations between these two systems.

The colonial declensionist environmental narrative, as defined by Diana K. Davis, has two essential parts: first, the belief that desert regions were once productive systems (often forests), and second, that their current state of degradation was caused by exploitation, especially overgrazing, from local populations.[9] Both ideas are present in desertification material produced by the UN, and the 1977 edition of the *Courier* is especially illuminating in this regard. Soviet scientist Victor A. Kovda's contribution, "World Trends in Aridification: A precarious balance upset," serves to introduce the issue and its importance to the magazine's readership. One of his illustrative examples is the loss of the "famous gum arabic acacia plantations" in Sudan, which, at the time, were suffering under conditions of drought and the depletion of soil nutrients—in other words, they were falling prey to "desertification."[10] He explains that the gum arabic trees are an example of desertification as caused by "[p]rimitive methods of farming, based on fallowing and the one-crop system, coupled with the age-old practice of excessive pasturing."[11] This is the same reasoning that Davis has identified in the colonial governments of Algeria and the Maghreb region, wherein perceived degradation was blamed on Indigenous populations living in these environments. Kovda similarly reproduces the belief that deserts were once green, productive systems when he compares the arid conditions of the 1970s to their "ancient bounty" as evidence of the land's deterioration.[12] Importantly, Kovda's article discusses very real and pressing issues of land degradation in the 1970s, which was impacting the production of gum arabic in Sudan, but his choice to frame this degradation within a colonial story of desertification shapes how these conditions are understood, and who is blamed.

In the context of colonialism, the idea that traditional land-use practices exploited the land and actually caused arid or desert-like conditions served as perfect justification for those arguing for greater colonial land control, which would ostensibly "help" these degraded lands return to their productive state. It is important to be clear here that traditional land-use practices like nomadic pastoralism are not inherently destructive, especially in comparison to the sedentary practices common to colonial settlements. Further, no land-use method is responsible for the creation of deserts, which are naturally occurring and lively ecosystems. Both ideas function to blame local people for the arid conditions in which they live, which further supported colonial policies and laws that removed land from Indigenous peoples and forced the implementation of European land management practices, especially sedentarization, even when such practices frequently resulted in degradation.[13]

The destructive potential of colonial land-use practices was not unknown at the time. For example, one of Kovda's colleagues in the 1977 *Courier* issue takes a very different view of the Sudanese trees. In Egyptian botanist Mohammad Kassas's contribution, "Are deserts man-made?" he points out that while the traditional land-use practices of nomadic peoples in Sudan facilitated the tapping of gum arabic trees for centuries (which is why they are "famous," to Kovda), the intensification of production in recent years was quickly leading to "desertification."[14] Nomadic cultivation, Kassas explains, followed cycles of migration and vegetation growth which included time for the land to recover and, perhaps most crucially, allowed for greater flexibility in response to environmental changes (such as periods of drought).[15] The trees were tapped and the gum arabic harvested on years when this was possible. By contrast, Kassas points out that the intensification of production brought on by increased population pressures in the region had resulted in overgrazing, over-production, and soil degradation. Kassas concludes that "[w]ithout a doubt, desertification is but the end result of ecological, economic and political pressures."[16] Although Kassas does not explicitly tie these changes to colonialism, the intensification of crop and livestock production was a hallmark of colonial governments across Africa.

At first glance, Kassas and Kovda's differing opinions are exemplary of the variation that has always existed within desertification discourse in the UN, which is not static and has been taken up differently by many individuals and organizations. But in the end, Kassas's critique of intensified land-use patterns resulting from current political and economic pressures does not disrupt the condemnation of nomadic practices by the majority of contributors to the 1977 issue. And, crucially, while many factors contributing to desertification were discussed at the UNCOD, rural populations are identified as the central focus of reform efforts. Although the narrative of desertification is flexible enough to contain the contradictions of Kassas and Kovda's competing ideas, they are not held in equal regard.

The disparity of Kassas and Kovda's ideas are also held together by their shared overarching goal, which is to demonstrate the need for more scientific land management practices to develop dryland regions. This focus on scientific and technological approaches to desertification serves to distance the UN's work on arid lands from its colonial context. But this distancing is superficial, relying on an idea of science as a politically neutral enterprise, which it is not. Kovda's article may not directly argue for colonial-style sedentarization policies, but his vilification of traditions like pastoral nomadism indirectly functions as a call for the implementation of a different way of living in the desert, one based on Western scientific knowledge. In 1977, the conquest of the desert is to be done by science through the integration of technologies such as irrigation and canal systems to transport water into drylands.[17]

The path from colonial narrative to UN science is quite direct. Prior to the UNCOD in 1977, the UN's interest in deserts was exemplified in UNESCO's Arid Zone Program, which ran from 1951 to 1962. The program aimed to improve scientific understanding of arid lands in order to aid in their development. It was active throughout much of the period of African decolonization and functioned as a new avenue for many colonial experts to take their work, and, consequently, their ideas of desertification.[18]

For example, the conception of desertification as a moving, threatening desert emerged from colonial researchers like British forest scientist E.P. Stebbing, who directly participated in the UN's Arid Zone Program. Stebbing helped to popularize the idea of desert advance in a 1935 article which detailed his experiences and observations in Northern Africa. This article (and others like it) helped to form the scientific basis of desertification. Years after his journey, in 1952, Stebbing prepared another report on the issue, entitled "Man-made arid zones," for UNESCO. Here, Stebbing reiterates the main findings of his oft-cited 1935 article, including his evocative depiction of the "encroaching Sahara."[19] In 1952, Stebbing was 80, a professor emeritus at the University of Edinburgh, and an old expert in the field. His report relies on familiar names and science—for instance, he cites Lavauden in his discussion of the past forest density in the Sahara region.[20] Stebbing and his 1952 report exemplify how the colonial ideas of a previous generation were taken up by the UN in the postwar period and integrated into that system of knowledge. Kovda's use of colonial notions of environmental crisis in 1977 therefore drew on what was, by then, well-established UN science.

Stebbing's ideas of desert advance are evident throughout the special 1977 *Courier* issue, which was entitled, "A Halt to Desert Advance" (see Figure 9.1). While the ten articles within present quite a variety of views on the topic, none stray far from this metaphor. The articles, including "Desertification: Climate as Accomplice," "A Winning Battle Against Destruction: Desert science is turning the steppes of Central Asia Green," and "Onslaught on the Nile: Shifting dunes threaten Egypt's breadbasket," sound more suited to wartime newspaper headlines than a scientific magazine, and they reveal the centrality of encroachment and threat to the UN's presentation of desertification.

The presence of the colonial desertification narrative in the 1977 *Courier* is especially important because of its function as promotional material for the upcoming

FIGURE 9.1 Cover page of the July 1977 edition of the *UNESCO Courier*, entitled "A Halt to Desert Advance."

Reproduced by permission of UNESCO.

conference. The issue was released in June 1977, just a few months prior to the conference in Nairobi, and as such, most of the contributors wrote on materials that were included in the proceedings. The ideas contained in Kovda's article and implicit in the titles of other contributions laid the groundwork for the 1977 conference, where desertification would be codified as an environmental crisis to rally around and an urgent issue for governments and international agencies to address.

Desertification as a Global Environmental Crisis at the UNCOD

UNESCO's Arid Zone Program shows us that the UN's interest in drylands and even in desertification has a long history, but the UN General Assembly didn't take a particular interest in the issue until the Sahelian Drought of 1968–1973.[21] Sometimes known as the Great Sahelian Drought, this crisis garnered worldwide attention as years of inadequate rainfall resulted in mass famines and crop failures across the region south of the Sahara. Severe drought is a natural part of arid and semi-arid systems, but the Sahelian Drought was particularly devastating due to the compounding pressures of increased populations in regions heavily impacted, socio-economic distress, and, relatedly, increased land use.[22] At the time, the severity of the drought was attributed to desertification, and although this correlation has been largely disproven, it spurred the UN to take a stance on the issue as part of its official response to the crisis.[23] And so the Conference on Desertification (UNCOD) sits between these two scales—participants sought to develop a globalized approach to the issue that would also function as the UN's response to the specific devastation of the Sahelian Drought crisis. The result was the institutionalization of desertification as a global environmental crisis, which relied on a simplistic story of deserts as an encroaching threat. This idea of desertification as "the spread of desert conditions," as it is consistently referred to in the conference report, follows from the conceptions of colonial scientists writing on the encroaching Sahara and results in a crisis narrative that is communicative, but which fails to capture the multi-storied complexity of desert spaces in times of crisis.[24]

This idea is reflected in several statements made by conference attendees included in the conference report, including that of Julius Kiano, President of the conference and Minister for Water Development in Kenya at the time. Kiano frames the work of the UNCOD as "the battle against the spread of deserts" and, in his closing remarks, explains his hopes that "the Conference could become another milestone in the history of man's conquest over the cruelties of nature and an event that historians would describe as having accelerated the reversal of desertification and promoted the encroachment of good lands upon desert areas."[25] This vision for the conference offers a striking portrayal of desertification as an enemy threat, but it also demonstrates the extension of this threat to deserts themselves. Deserts are positioned in Kiano's remarks as the opposite of "good lands," regardless of whether they are degraded, in a state of drought, or actively desertifying. Statements like these easily translate into fear of the menacing, encroaching desert, and promote a singular view of land degradation and drought in arid and semi-arid lands that is based on notions of scarcity and lack, thereby obscuring other stories of deserts that might imagine crisis differently.

The way that the report and conference materials portray desertification is important because establishing the term as a field of scientific inquiry was one of the main goals for the event. Prior to the conference itself, extensive work was done to collect information on the drought and land degradation in drylands, including a series of preparatory meetings and the creation of case studies, maps, and other data prepared through consultation with various scientists and specialists on drylands (as well as much of the research from the old Arid Zone Program). This imagery of battle against hungry desert lands seems out of place in a report that is seeking to establish desertification as a valid field of scientific inquiry, but this language is crucial for the formation of a compelling narrative of environmental crisis. And indeed, many participants understood the formation of an effective narrative to be the ultimate goal of the UNCOD.

In the UNESCO archives, there are a series of audio interviews that were recorded in the lead-up to the 1977 event.[26] In these interviews, scientists, politicians, and ambassadors are asked to give their perspective on the conference and what they hope it will achieve. In one interview, British journalist Jon Tinker, who also contributed to the preparation of the event, describes the aim of the conference as "an exercise in the concentration of political will."[27] Laying out the facts of desertification (such as they were) and establishing a plan of action weren't *really* the point of the conference; as Tinker explains, much more important was to present the ideas in such a way as to inspire international recognition and action.[28] Certainly, the development of concrete plans like the Plan of Action to Combat Desertification (PACD) was meant to provide guidelines to fight desertification and make real change, but the words of these policy leaders and the convention text itself suggest that this direct outcome is secondary to the ideological emphasis of the conference, which was to establish desertification as a global concern deserving the attention and support of the UN and its associated agencies.

Establishing desertification as a global environmental crisis was no easy task. While the Sahelian Drought was clearly a crisis event, the idea of a global environmental crisis was not well established. Today we are intimately familiar with this idea in the context of climate change, but the concept was far from popular in 1977. In fact, geographer Nick Middleton suggests that desertification may be the first globally recognized environmental crisis, a designation that is based on the work done to establish the concept at the UNCOD.[29] Therefore, the narrative work that the UN was doing in the 1970s should be considered formative, both for the general idea of environmental crisis and for the particular representational challenges of severe drought and land degradation.

Both drought and soil degradation are examples of what Rob Nixon has called slow violence, in that their variability makes them particularly difficult to recognize and talk about.[30] Drought, for example, is common in drylands, it occurs frequently and its impacts vary greatly across any given area, only becoming a crisis when it is particularly severe or unrelenting. Similarly, soil degradation is fairly obvious when severe, but otherwise requires careful study to notice changes over time. Even then, a degraded landscape can be hard to recognize. A field of crops may look fine on the surface while the soil is losing nutrients at a rate that will eventually lead to lower crop yields. These characteristics are entirely ill-suited to the spectacle-driven media and news cycle which has only become more exaggerated since Nixon's writing in 2011. The UN's narrative of desertification functions as a transformation of the otherwise slow and uneven violences of land degradation and drought into an environmental crisis more befitting the cinematic, newsworthy catastrophes that surround us. In other words, desertification is a more powerful story.

Unlike drought and land degradation, the idea of desertification as the threatening spread of desert conditions addresses many of Nixon's representational obstacles. Compared to the complicated scenarios of degradation and drought, desertification is easily explained and pictured: "deserts on the march" can readily be understood as dunes rolling over landscapes and fields blowing away in the wind. These ideas offer potent visuals that effectively communicate a sense of urgency and threat, and they do so more loudly than a semi-arid shrubland that has a decreased chance of surviving the next drought cycle.

However, the potency of its visuals does not mean that desertification, as a narrative of crisis, is a success. Nixon was calling for a re-conceptualization of these issues that could capture the full complexity of slow violence, and that is not what desertification does. Rather, desertification offers a *visually compelling misreading of land degradation*, a poor translation that is perhaps easier to grasp and certainly powerful to picture, but which is harmful in its simplification of the realities of arid lands in times of crisis. Kiano's remarks, which paint all desert areas as a threat, offer one example of the harm that comes from desertification narratives.

This tradeoff between effective storytelling and inaccuracy was not unknown to the conference participants. In another audio interview, this one conducted in January 1977, during one of many preparatory meetings, J.A. Mabbutt, a consultant with the UNEP from the University of New South Wales who worked on the case studies and overview documents, was asked whether he thought the phrase "deserts on the march" was a useful picture for people to have when considering desertification. Mabbutt responded that it was, at least in some areas and geographies, though he cautioned that this picture "should not conceal the nature of the extension of these desert conditions"—which is to say, deserts do not actually march over new territories.[31] "In fact," Mabbutt elaborates,

> the progress of desertification is rarely a wave-like front or advance … by far the commonest circumstance is the sporadic occurrence, almost spot-like, of denuded lands [ ] in relatively vulnerable areas in the desert margin[s], and these areas tend to extend and to link up and become further degraded as the process of degradation continues.[32]

Mabbutt's response acknowledges that the idea of the advancing desert is misleading (even inaccurate), but still useful. And if your goal is to spread the word about desertification as a concept and idea, then Mabbutt is probably right, just as Kiano's representation of desert lands as threats, while perpetuating the villainization of deserts, was also an effective means to garner much-needed support for drought relief. However, the idea of advancing desert margins does, as Mabbutt warns, conceal the nature of the problem in public understanding. And in fact, more than concealing the scientific realities of land degradation, the simplification of this narrative obscures the complex histories and socioeconomic causes of desertification as a colonial narrative construction. The metaphor of desert advance is unable to account for the history and detail of how land is used in arid regions—it is a single-storied desert, only able to communicate a sense of crisis and spectacle.

And yet, the ability for a simplified, metaphor-heavy idea of desertification to communicate a sense of urgency and crisis is exactly why Mabbutt couldn't deny the utility of ideas like "deserts on the march" and must be understood as part of the conference's success. The organization of the UNCOD had an immediate effect on writing about the spread of soil issues. One example can be seen in the work of US ecologist Paul B. Sears, who, at the time of the conference, was working on his 1980 edition of *Deserts on the March*, a text initially published in 1935 to make scientific and ecological sense of the Dust Bowl disaster.[33] While the title of the text makes it clear that Sears was dealing with topics central to desertification, the term does not appear in any of the earlier editions. Despite its long history, it was not widely known or utilized until the UNCOD. In the fourth edition of his book, Sears repeatedly mentions the 1977 conference in Nairobi and expresses his hope that the issue of desert advance will finally be addressed now that international governments were gathering to discuss the issue and had named it *desertification*, which he felt directly attributed the issue to human action.[34] The fact that the UN conference did not come up with the term didn't matter—Sears's text exemplifies the way in which the UNCOD effectively established desertification as a global concern, at least in the eyes of the public.

This success has been long lasting, considering that while the contributors to Behnke and Mortimore's 2016 collection, *The End of Desertification?* tear down its science with relative ease, it is the institutionalization of the term that they are unable to effectively solve.[35] The 1977 conference was not able to agree on a world map of desertification or even a stable definition for the phenomenon, but it was able to concretize this particular narrative of land degradation as a moving desert front. It brought together the work of UNESCO's Arid Zone Program and all of the colonial science involved and translated everything into a narrative of environmental crisis for the new age of environmentalism.

The UNCOD laid the groundwork for the public understanding of desertification that we have today, which adapts many aspects of the colonial narrative for use as spectacle. The idea of desertification as a conflict between encroaching deserts and humanity, specifically rural inhabitants of dryland regions, provided a useful crisis with which to engage the international community. In 1991, a report was published by the UNEP that condemned the scientific results of the UNCOD as a failure, but it said nothing of the story which the conference established. The narrative of desertification that emerged from the conference, its characterizations of the desert and desert peoples, as well as its prevailing metaphors, entered public knowledge and was quickly adapted to the growing concerns about global anthropogenic climate change.

Continuing the Battle with the UNCCD

The 1996 UN Convention to Combat Desertification (UNCCD) was designed to address the failings of the 1977 conference, and from a practical standpoint, the associated texts do appear to be quite different. The convention text includes no reports, discussions, or other evidence of the people behind the policies contained, which means that there are no metaphorical descriptions of personified desert threats. Instead, the convention is streamlined, composed of a short text outlining the issue and the goals to be achieved. To replace the more general PACD that emerged from the UNCOD, the convention is based around five regional implementation annexes that attempt to guide in the planning and financing of efforts to mitigate land degradation in areas considered at high risk of desertification and/or severe drought. The use of regional implementation annexes was meant to solve the contradiction of a convention that specifically focused on Africa while still positioning desertification as a global problem.[36] But the annexes are also indicative of policy that is more specific and immediately useful.[37] In comparison to the UNCOD, the 1996 convention was meant to be more realistic, more scientific, and to end in measurable results. However, despite the many attempts to resolve the issues of its predecessor, the convention text and surrounding materials reveal the preservation of the narrative of desertification that was codified in 1977. The UNCCD is designed to be a neutral, scientifically informed text, but even so it is telling a story, and this story has real implications for how we approach land degradation and drought in drylands.

The convention text spends no time debating the definition that eluded those involved with the UNCOD, and instead opens with a brief description of desertification as "[l]and degradation in arid, semi-arid and dry sub-humid areas resulting from various factors, including climatic variations and human activities."[38] This definition still stands today. It is general and somewhat vague, but it works in that it is widely applicable to a variety of soil concerns—and that's exactly what it was designed to do.

The UNCCD was a result of the 1992 UN Conference on Environment and Development, also known as the Rio Earth Summit. One of the biggest obstacles to the approval of the UNCCD at Rio was, perhaps unsurprisingly, coming up with a solid definition. Bo Kjellén, who chaired the working group dealing with desertification at Rio (and then chaired the International Negotiating Committee on Desertification that was tasked with creating the convention itself), explains in a personal reflection that the question of desertification's definition "had haunted the international community for years, particularly since it was linked to scientific controversies about the real nature and extent of the problem."[39] His solution at the Rio Conference was simple and effective: to avoid the issue. As he explains:

> Of course I realized that [defining desertification] was an important question, but I did not feel that it should be allowed to block progress towards real action, and therefore encouraged delegates to try to find a pragmatic solution that would be broad enough to cover all the central aspects of land degradation in the drylands. And I continue to feel that the outcome was satisfactory, since the Agenda 21 definition of desertification has never been seriously challenged and has enabled the international community to make interpretations which have facilitated concrete action in the field in many different parts of the world.[40]

And so, the UNCCD's definition of desertification is not intended to be necessarily definitive or even accurate: it is intended to be inoffensive and unobtrusive to the functioning of the convention and the idea of desertification. This mirrors the outcomes of the UNCOD, in that the institutional utility of desertification as an idea and story remains more important than the question of its definition and accuracy.

This also helps to explain why the 1996 convention was built around desertification at all, rather than a more accurate or less contentious alternative. The term is still useful as a word that is packed with cultural significance: desertification, and not land degradation, is associated with a compelling story of crisis while retaining Eurocentric biases related to land use. The convention text mentions desertification a total of 129 times across its 54 pages. In every instance the word could be replaced with the phrase "land degradation" and the meaning of the sentence would not change.[41] This theoretical replacement would free the UNCCD from the scientific controversy that was already building around the topic in the 1990s, but it would also change the cultural resonance of the convention. The usefulness of desertification for this document and the convention itself exists outside of its generalized definition and is tied to the cultural narratives associated with the term.

Just as holding onto the term itself is indicative that the convention upholds the narrative established by its predecessor, the phrasing of the convention's title is also significant. The 1992 Rio Earth Summit approved the creation of three landmark conventions focusing on environmental concerns: the UNCCD, the UN Convention Framework on Climate Change (UNFCC) and the UN Convention on Biological Diversity (UNCBD). Of the three conventions, only the UNCCD includes a transitive verb, *combat*, in its title. Compared to the descriptive titles of its sister conventions, the UNCCD positions its focus in the context of conflict and confrontation. Not only does this set it apart from the other Rio conventions, but it is also highly instructive as to why desertification is useful for the convention at all—because this suggests that its focus is perhaps not desertification itself, or not *only* desertification but rather the particular relationship of combatting desertification. The focus on conflict makes sense as a reference to the UNCOD's failed PACD—but it also demonstrates that while the new convention was designed to address many of the scientific and conceptual failings of its predecessor, it maintains the old narrative of desertification and places humanity in an oppositional relationship to a vilified image of desert lands. This narrative continues to offer a misreading of land degradation and drought in arid and semi-arid regions, which directly shapes imagined responses to the issue.

For example, the often-stated goal of the UNCCD is Land Degradation Neutrality (LDN), wherein the spread of degraded areas would cease to grow as it does now. If we continue to think using the terms of conflict established by the UN's desertification narrative, LDN could be seen as an end to hostilities with the desert, where land degradation (the central conflict) still exists, but we are no longer in an active state of war, and the desert is no longer "invading." LDN suggests that the goal of the UNCCD is fundamentally neutral rather than positive. A positive solution might involve land-use practices that are regenerative, which add nutrients to the soil over time, or a greater commitment to de-centralized land control. LDN, in contrast, is more like the *management* of degradation, or the maintenance of neutrality. It exemplifies the way that crisis narratives can result in the stabilization of present conditions.

Anthropologist Joseph Masco argues that the term crisis has become a stabilizing force that serves to manage disaster conditions rather than enact real and even revolutionary change. Stabilizing ideas, like LDN, fit more easily into the UN's narrative than more radical approaches that might prioritize the rights and needs of local farmers and communities, or reconcile with the complex colonial history of ideas like desertification. Indeed, Masco also points out that the feverish focus on the emergencies of the present moment allows crisis writing, discourse, and narratives to avoid a meaningful engagement with historical precedent. He writes that "[c]risis talk today seeks to stabilize an institution, practice, or reality rather than interrogate the historical conditions of possibility for that endangerment to occur."[42] This helps to explain how the UN's narrative of desertification can maintain so many of the dominant aspects of the colonial story without ever coming to terms with this history. There is simply no place for the history of desertification and its colonial entanglements in the UNCCD's framework, and therefore in the larger narrative of desertification supported by the UN.

This avoidance of colonial contexts is apparent in UN publications like Carole Douglis's *Trishna and the Dream of Water* (2006). This book is part of the TUNZA Environmental Series for Children, designed to educate youth on issues such as melting polar ice, plastic pollution, and deforestation. In it, Douglis tells the story of a young girl, Trishna, who receives advice from the wind, sun, earth, and water on how to treat the land to increase the amount of water that's readily available. By the end of the story, Trishna inspires others to follow this advice and regreen their dryland community. Trishna's plight contains many aspects of the colonial desertification narrative, including the idea of desert advance (as the dunes near her town get closer every year) and the belief in past productivity.[43] Trishna is told by each of her elemental guides that she must learn to treat the land as her ancestors did, indicating that the goal is to *return* her community to its former, greener state.[44] The present degradation that Trishna's community experiences is therefore the result of human action, and essentially, a failure to use the land well.

The question sitting at the heart of the story is, of course, what exactly could have happened for the people of Trishna's community to "forget" all of their traditional methods for living on the land. What drove the young people of her village to move to the city for work instead of staying to care for their community? While a children's book is perhaps ill equipped to deal with all of the complexities of postcolonial life, that doesn't erase the harm of their absence. This generational gap in the story represents the erasure of the historical conditions that led to Trishna's hardship. This history is not seen as important to the story of desertification. Instead, the book focuses on the hard work of the community in finally utilizing the right tools and knowledge to improve their lives and their land.[45]

There is more at stake here than just a recognition of historical causation. As Masco points out, the inability of crisis narratives to reckon with the past also limits their ability to imagine new and radical futures. A decolonial approach to land degradation is impossible for the UN unless the colonial histories embedded in the idea of desertification are recognized and meaningfully dealt with. Until then, the solutions offered and supported by the UNCCD are forced to fit into its narrow narrative framework. Even as the UNCCD is more scientifically careful or accurate, it has not adjusted the central narrative of desertification on which the UN has always relied. Therefore, the convention remains as limited in terms of possible solutions and responses to land degradation as the initial UNCOD, and indeed any colonial response to land degradation.

Looking Ahead: What Do We Do with Desertification?

The United Nation's story of desertification aids in a critical consideration of what it means to narrate environmental crisis. These stories are an intricate part of our collective search for ways through the climate emergency. They shape how we understand what is happening to the world, as well as why it has happened, and how we imagine ways to move forward. Since the 1970s, desertification has been, in some ways, an effective method for the UN and its associated agencies to discuss the idea of land degradation and drought in arid lands. At the same time, desertification is an idea that has enabled the misrepresentation of land degradation as a conflict between humanity and desert lands. The striking and visual conceptualization of desertification as a moving desert front has enabled a broad public recognition of the concept at the cost of cementing the harmful association of deserts with degradation. If we think of desertification as a way of garnering resources and support for issues that are otherwise slow and difficult to understand, then we can see its usefulness, but the reality of the situation is much more complex and troubling. It begs the question—what do we do with desertification?

Returning to Mark Stafford Smith's observation, it seems unlikely that we will be able to remove desertification from public discourse as long as the UN continues to use it. However, there is evidence that the UN's use of desertification is shifting. Geographer Nikolai Dronin has recently argued that the UNCCD leadership has generally accepted land degradation as a more scientifically accurate and geographically inclusive term.[46] This development is significant and, as Dronin argues, indicates the ability for the scientific community to shift political conversations. However, as I have demonstrated in this chapter, scientific evidence is not what makes desertification a popular and persistent idea; rather, its strength lies in its existence as a compelling cultural narrative. The UNCCD may be internally leaning toward a focus on land degradation, but its title remains unchanged. Perhaps more importantly, desertification remains a key term in highly publicized initiatives such as the 17 Sustainable Development Goals established by the UN General Assembly in 2015, and the IPCC 2019 special report, *Climate Change and Land*, which devotes chapters to both desertification and land degradation. In this regard, things appear to have changed very little since the UNCOD, where experts determined that desertification was an uncertain concept in need of further study even as the emphasis of the conference functioned to institutionalize the term as a popular narrative of environmental crisis.

Land degradation itself, although a more accurate term, is not a perfect solution. Crucially, land degradation may not be the most pressing concern for arid lands. Recent data has shown that land degradation itself is far more common outside of Earth's arid regions—and is especially concerning for areas of high agricultural production.[47] Fixing the scientific inaccuracies of desertification in the UN may only serve to move the discourse further from the needs of deserts and dryland peoples. Here it is important to remember that the UNCCD was conceptualized and fought for by representatives of African nations who felt that the UN's growing environmental efforts needed to pay more attention to the specific problems impacting Africa's drylands.[48] There is still a need for narratives that can help us to articulate and understand environmental crisis in arid and semi-arid lands, but desertification is not enough. We need a multitude of storied deserts through which we might capture the variation and complexity of arid and semi-arid lands in times of environmental crisis.

And so we return to our initial question—what do we do with desertification? If there is a future for desertification as a term, it must involve a reckoning with the idea as a neocolonial concept. Looking closely at the UN's narrative reveals the ways in which stories of environmental crisis are entangled in complex histories and institutionalized in structures of power and expertise. If we are to embrace new narratives and new ways of talking about land degradation in drylands, it must not come at the cost of erasure. Alternative terminology like land degradation more effectively addresses the scientific inaccuracies of desertification and the harmful misrepresentation of desert lands, but it does not make visible the neocolonial construction of the idea, especially in the dryland regions of Africa.

There is no easy answer to the question of desertification, but it is clear that any reimagining of the UN's approach to land degradation in arid lands will require a close understanding of how desertification is functioning as a cultural narrative as well as a scientific and political concept. More than anything, the development and utilization of desertification narratives in the UN demonstrates the need to stop fighting deserts and combatting desertification and begin to imagine stories that can help us to live and grow with drylands in times of crisis.

Notes

1 Smith, "Desertification: Reflections on the Mirage," 555.
2 Kjellén, "The Saga of the Convention to Combat Desertification," 128.
3 FAO, "Erosion, Drought and Deserts," 8.
4 Davis, "Deserts and Drylands Before the Age of Desertification," 204.
5 Lynch, "Foreword: Desertification," ix.
6 Mortimore, "Changing Paradigms for People-Centered Development in the Sahel," 65.
7 Thomas and Middleton, *Desertification: Exploding the Myth*, 1994. See also Selcer, *The Postwar Origins of the Global Environment,* 131.
8 Swift, "Desertification: Narratives, Winners, & Losers," 73–90.
9 Davis, *Resurrecting the Granary of Rome*.
10 Kovda, "World Trends in Aridification: A Precarious Balance Upset," 12.
11 Kovda, 12.
12 Kovda, 12.
13 Davis, *Resurrecting the Granaries of Rome*, 168; Toulmin and Brock, "Desertification in the Sahel," 37, 38. As geographers Camilla Toulmin and Karen Brock point out, anti-desertification efforts today continue to support policies based on large-scale, centralized approaches to land management, including human resettlement.
14 Kassas, "Are Deserts Man Made?" 6.
15 Kassas, 6.
16 Kassas, 6.
17 Babaev and Orlovsky, "A Winning Battle Against Destruction," 20.
18 Selcer, 64; Davis, *The Arid Lands,* 144–145.
19 Stebbing, "Man-Made Arid Zones," 4.
20 Stebbing, 4.
21 This is the date range for the Sahelian drought used for the UNCOD, but other sources give different ranges. Drought is difficult to measure, especially across such a large area. It is not a static phenomenon and conditions vary at local scales.
22 The Sahel is a transitionary ecoregion that follows the southern edge of the Sahara desert, intersecting portions of several nations, including Senegal, southern Mauritania, central Mali, southern Niger, northern Nigeria, central Chad and Sudan, Eritrea and parts of northern Ethiopia and South Sudan.
23 Modelling studies like that of Giannini (2016) have shown that the Sahelian Drought was largely caused by climatic factors and the extremes of variability that are natural in arid regions.
24 "Report of the United Nations Conference on Desertification," 3.
25 "Report of the United Nations Conference on Desertification," 79, 105.
26 The metadata on these audio files is often lacking—while a date of recording is usually given, the place or occasion is not. By UNESCO's own admission, the audio records on their website arc posted as-is from paper records and may contain errors. In my

research I found interviews that were incorrectly titled and listed the wrong interviewees. The name of the interviewer is never provided, although in some cases it is listed as the UNEP. Similarly, there is no data on where these recordings were distributed, but it was UNESCO Radio's policy at the time to make their programs freely available to any station that requested them.
27 Tinker, "DESERTIFICATION UNEP Interview with Don Tinker," 00:35:38–00:35:42, transcription by author.
28 Tinker, 00:35:43–00:36:04, transcription by author.
29 Nick Middleton, *Deserts: A Very Short Introduction*, 107. See also Selcer, *The Postwar Origins of the Global Environment* (2018). Selcer connects the UN's work on deserts and desertification to the development of a global conception of the environment.
30 Nixon, *Slow Violence and the Environmentalism of the Poor*, 2.
31 Mabbutt, "M. Daih Mabutt (Australia) on Desertification," 00:02:36–00:02:48, transcription by author.
32 Mabbutt, 00:02:49–00:04:10, transcription by author.
33 The Dust Bowl itself, as a catastrophic example of land degradation tied to human action, was frequently cited within the context of the UNCOD as a model for anti-desertification efforts based on the implementation of science and policy-based solutions. See, for example, soil scientist Harold Dregne's contribution to the 1977 *UNESCO Courier* issue, "The Sands of Wrath: America's Dust Bowl in Retrospect."
34 Sears, *Deserts on the March*, 252. Interestingly the conference did not agree that desertification was solely caused by human action and many attendees/reports suggested that a combination of climatic circumstances and human action was more likely. Sears's interpretation is representative of the public understanding of desertification as an anthropogenic environmental crisis that would emerge following the conference.
35 See Smith, "Desertification: Reflections on the Mirage," for a discussion on this topic.
36 The full title of the convention is: United Nations Convention to Combat Desertification in those Countries Experiencing Severe Drought and/or Desertification, Particularly in Africa.
37 Kjellén, 128.
38 *UNCCD*, 4.
39 Kjellén, 128.
40 Kjellén, 128.
41 *UNCCD*, 51, 53. This is only counting the convention text itself, if you include the regional annexes, there are two exceptions in Annex V for Central and Eastern Europe (which was adopted later, in December 2001) where desertification is listed twice *alongside* land degradation, indicating that they are separate ideas. Although the distinction between the two is never explained.
42 Masco, S73.
43 Douglis, *Trishna and the Dream of Water*, 4.
44 Douglis, 7–12.
45 Douglis, 17–18.
46 Dronin, "Reasons to Rename the UNCCD," 16.
47 Davis, *The Arid Lands*, 9–10; Dronin, 2.
48 Kjellén, 128.

Bibliography

"A Halt to Desert Advance." *The Unesco Courier: A Window Open on the World*, July 1977. p. 1. Catalogue number 0000074813, UNESCO Archives.

Davis, Diana K. "Deserts and Drylands Before the Age of Desertification." In *The End of Desertification?: Disputing Environmental Change in the Drylands*, edited by Roy Behnke and Michael Mortimore, 203–223. Springer, 2016.

Davis, Diana K. *Resurrecting the Granary of Rome*. Ohio University Press, 2007.

Davis, Diana K. *The Arid Lands: History, Power, Knowledge*. MIT Press, 2016.

Douglis, Carole. *Trishna and the Dream of Water*. Tunza Environmental Series for Children. United Nations Environment Programme, 2006.

Dronin, Nikolai. "Reasons to Rename the UNCCD: Review of Transformation of the Political Concept through the Influence of Science." *Environment, Development and Sustainability*, January 28, 2022.

Giannini, Alessandra. "40 Years of Climate Modeling: The Causes of Late-20th Century Drought in the Sahel." In *The End of Desertification?: Disputing Environmental Change in the Drylands*, edited by Roy Behnke and Michael Mortimore, 265–291. Springer, 2016.

Kjellén, Bo. "The Saga of the Convention to Combat Desertification: The Rio/Johannesburg Process and the Global Responsibility for the Drylands." *Review of European Community & International Environmental Law* 12, no. 2 (July 2003): 127–132.

Lynch, Tom. "Foreword: Desertification." In *Reading Aridity in Western American Literature*, edited by Jada Ach and Gary Reger, ix–x. Lexington Books, 2020.

Mabbutt, J. A. M. Daih Mabutt (Australia) on Desertification: (UNEP Programme). Audio Recording, January 3, 1977. UNESCO Digital Archives.

Masco, Joseph. "The Crisis in Crisis." *Current Anthropology* 58, no. S15 (February 2017): S65–S76.

Mortimore, Michael. "Changing Paradigms for People-Centered Development in the Sahel." In *The End of Desertification?: Disputing Environmental Change in the Drylands*, edited by Roy Behnke and Michael Mortimore, 65–98. Springer, 2016.

Nixon, Rob. *Slow Violence and the Environmentalism of the Poor*. Harvard University Press, 2011.

"Report of the United Nations Conference on Desertification." *UNEP*, 1977.

Selcer, Perrin. *The Postwar Origins of the Global Environment: How the United Nations Built Spaceship Earth*. Columbia University Press, 2018.

Smith, Mark Stafford. "Desertification: Reflections on the Mirage." In *The End of Desertification?: Disputing Environmental Change in the Drylands*, edited by Roy Behnke and Michael Mortimore, 539–560. Springer, 2016.

Stebbing, E. P. "Man-Made Arid Zones." UNESCO Digital Archives, November 24, 1952.

Swift, Jeremy. "Desertification: Narratives, Winners, & Losers." In *The Lie of the Land: Challenging Received Wisdom on the African Environment*, edited by Melissa Leach and Robin Mearns, 73–90. The International African Institute, 1996.

Thomas, David S. G., and Nicholas J. Middleton. *Desertification: Exploding the Myth*. John Wiley & Sons, 1994.

Tinker, Jon. 'Desertification' UNEP Interview with Don Tinker. Audio Recording, May 20, 1977. UNESCO Digital Archives.

Toulmin, Camilla, and Karen Brock. "Desertification in the Sahel: Local Practice Meets Global Narrative." In *The End of Desertification?: Disputing Environmental Change in the Drylands*, edited by Roy Behnke and Michael Mortimore, 37–63. Springer, 2016.

"United Nations Convention to Combat Desertification in Those Countries Experiencing Serious Drought and/or Desertification, Particularly in Africa." UNCCD, 1994.

10

THE ATACAMA DESERT

Environmental Imagery from Presumably Void Geographies Between the Andes Mountains and the Pacific Ocean Cliffs[1]

Andrea Casals-Hill and Pablo Chiuminatto

Milestones

Those of us growing up and being schooled in the 1970s and 1980s in Chile learned by heart that Chile is a long and narrow country, spanning from the driest desert in the world all the way south, including Antarctica. The superlative "driest" for the Atacama Desert was a strange source of nationalist pleasure. Several studies have affirmed that it is the driest desert in the world with the most extreme conditions; so much so that it is compared to Mars (Azua-Bustos et al., 2022).

There still is a kind of chauvinism in discourses of Atacama, where it is essential to understand that a large share of that desert was integrated after the bloody war called *Guerra del Pacífico*, 1879–1884. Historically, however, we should call it the "saltpeter war," since the expansionist motivation of the relatively new nation was to incorporate vast sections of that desert that held the largest deposits of the non-metallic mineral, nitrate, as well as huge copper deposits (Barnum 2003).

Though the desert appears to be a vacuum to strangers, at the same time it turns out to be a permanent site of disputes mainly because of the minerals that constitute its geology: yesterday it was saltpeter, after it, it was copper, and now lithium. But it is not only the solid ground and underground that are coveted; in the last 50 years its starry nights taking place in the highlands of *Los Andes* mountain range have been regarded as among the most pristine nocturnal skies in the world (Doyle 2021). Moreover, towards the west, the Atacama Desert borders with the Pacific Ocean and, therefore, with another source of abundance: that immense ocean, with all its life forms under the rushing waters and upwelling currents.

The Atacama Desert covers an area of approximately 105.000 km^2 (65,244 miles), it is almost 1,600 km (994 miles) long and has a maximum width of 180 km (112 miles) across latitude 52°21′ south, which is also the widest milestone

DOI: 10.4324/9781003397212-14

in the country. As mentioned before, it is bordered by the Pacific Ocean to the west and by the Andes Mountain range to the east. According to specialized studies, the Atacama Desert has had its current characteristics and the existing climatic features for about 8,000 to 5,000 years before the present, when a great drought occurred (Garreaud et al. 2010). This drought broke the equilibrium of what was a humid ecosystem "where camelids and associated plant resources were concentrated" (Nuñez et al. 2010, 364), and human settlements even developed pastoralism. It is in fact a complex process to account for, yet experts have proven that the area was populated for over 13,000 years. In the words of Santoro et al.: "Given that environmental and social systems are in constant interaction, our [research] model focuses on the interaction between ecosystem services and adaptive strategies to describe socio-environmental continuities and discontinuities" (2017, 28) of the Atacama Desert.

On the one hand, the coastal mountain range is an actual boundary, a natural topographical barrier that prevents the entry of humidity coming from the Pacific Ocean, from the west towards the desert plains. On the other hand, and towards the east, the Andes Mountain range prevents the humid atmospheric systems coming from the Amazonian areas to reach the desert. Thus, the Atacama Desert *per se* is a geographical barrier at our northern frontier, that at the same time is framed by the mountains, both east and west, contributing, at present, to its superlative aridness.

Paradoxically, at the southern end of our long and narrow country lies the planet's actual largest desert: Antarctica, defying our stereotypical images of deserts as hot and bone-dry. Featured in *National Geographic*'s February 2023 issue, Antarctica meets the key criterion for such a consideration: minimal rainfall. Places like McMurdo Dry Valleys have received no rain in 14 million years. Nonetheless, both icy Antarctica and arid deserts like Atacama can evoke a similar sense of vast emptiness (Rodríguez 2023). Hector Rodriguez writes: "our collective imaginary … responds to the idea of a warm and dry place, hit by burning sun during the day, relentless wind, and falling temperatures during the night."[2]

Whether icebound or dry cracked plains, the imaginaries that these natural frontiers tend to elicit revolve around vast emptiness, and Chile's literary and cultural imaginary has been set up between these two deserts, north and south. We may appreciate these imaginaries eloquently in literary productions and in visual arts, in a process that moves from the natural to the cultural (Sagredo 2014). But to speak of Chile's deserts is not only to think of a bioregion in the geographical sense; it is also to speak of multiple spaces and overlapping layers: economic, political, cultural, and environmental. Likewise, there are various lenses through which a selection of works and authors can be summoned to explore and describe the diverse stories that the desert prompts.

One of the aspects that contributes to constructing the desert as an estranged landscape is precisely the assumption that it is a void, even though we know of the different dimensions of life and history that have been present in it for thousands of years. Each of the elements that we can identify as belonging to and characterizing the Atacama Desert is reflected not only in the current reading of its horizons but also in the ancestral comprehension of it. What we see portrayed as vestiges when we travel through the Atacama Desert are evidence of the ancestral settlements, together with remains from the saltpeter extraction period (complete company towns that were abandoned over a century ago), as well as new human settlements and production sites related to contemporary mining processes. Atacama reveals the coexistence of time frames that go back to archeological and mythical all the way up to the anthropo-scenic. As Victoria Castro rightly points out:

> The camelid [guanaco, alpaca, vicuña] is not only a food or technological resource, but much more, it populates the terrestrial world and the celestial world, and in this sense, we can assert that it becomes part of the very life of the Andean people and communities in the profane and the divine as a life-producing force. [...] In this Andean world that it mirrors, you will see many other non-human beings that populate the celestial sphere and many others that accompany and satisfy the human being.[3]
>
> *El camélido no es sólo un recurso alimenticio o tecnológico es mucho más, puebla el mundo terrestre y el mundo celeste y en ese sentido podemos aseverar que se inserta en la vida misma de las personas y las comunidades andinas en lo profano y en lo divino en tanto fuerza productora de vida ... En este mundo andino que espejea, verás muchos otros seres no humanos que pueblan la esfera celeste y tantos otros que satisfacen al ser humano.*
>
> (Castro 2022, 94)

A recent version of this mirroring can be found in the selection of literary works that we would like to bring together to reappraise this area, but this time from the prism of anticipation. Atacama Desert's deep time story shows us how the desertification of the last millennium becomes another mirror, though this time it is one related to the Anthropocene (Gayo et al. 2019; Workman 2020). We dare to say that to a certain extent, the Atacama Desert is the prolepsis of the ecological catastrophe that we may extrapolate to the changes we see in the current global climate conditions.

As we research to write an ecocritical story of the Atacama Desert, we notice that most of the stories about this desert are or have been told from the perspective of male authors, historians or storytellers, male narrators, male poets and even male characters, constructing an epic of domination and extractivism. We propose that looking at this storied desert through Ursula K. Le Guin's *The Carrier Bag Theory of Fiction* (2019) is a fruitful exercise that can broaden our understanding of this bio-geo-political-region. In her essay, following Elizabeth Fisher's assertion that the first cultural artifact must have been some sort of container to keep or transport the food that was gathered, Le Guin wonders why the stories told are about the male hunters with spears, sticks, bones, and swords; how did stories about heroic male hunters prevail over stories of foraging and gathering. Even the novel, where heroes take over the scene, has a linear development). Thus, Le Guin opens our appreciation of the stories told about Atacama. To the sequence of spears and swords that Le Guin foregrounds in her essay, we would like to add the miner's pillory and the writer's stylus. We do not intend to impose a feminist reading of the desert, but rather to try to revisit the epic traces that mark this territory's stories and history. In such stories, the *carrier bag* is missing. Undoubtedly, woven bags were used by the archaic peoples of the desert, like the *Chinchorro* culture (7020–1500 a.C.) that lived from fishing and wove fishing nets (*chinguillo*) and baskets (Standen and Arriaza 2014; Montt et al. 2023). These artifacts helped them populate the desert, blurring the traces of the masculine epic story that we observe at first sight in contemporary narrations of the desert.

Atacama's official history and traditional timeline speak of conquest (first the Incas as predecessors, then the Spanish); of military wars, particularly the war for the territory where saltpeter lays; of the extractivism at the largest open pit copper mine; of the social struggles of the near enslaved workers in the early XX century; of the prison camps under the dictatorship in the 1970s and 1980s. All these are epic stories, journeys or campaigns deployed, have been registered and told from the perspective of almost exclusively male voices (Figure 10.1). This fate is exemplified by Salvador Reyes, a writer from northern Chile, in many of his short essays in the book *Andanzas por el desierto de Atacama* (trad. Wanderings through the Atacama Desert), published in 1969.

FIGURE 10.1 Atacama desert, saltpeter "pampa."

Photo Patricio Salinas in Atacama Captivity Geometry (2020).

The Poet's Desert

In the poetry collection *Purgatorio* (1979),[4] Raúl Zurita (1950) allows both the infinite quality of the Atacama Desert (naming it), and the (many) deserts' sterility to move on.

i Let the infinity of the Atacama Desert pass
ii Let the sterility of these deserts pass [....][5]

i *Dejemos pasar el infinito del desierto de Atacama*
ii *Dejemos pasar la esterilidad de estos desiertos [....]*[6]

For Zurita, as for many of us, the infinite quality seems to be an asset that awards the Atacama Desert to be signaled and named, while the perceived bareness of the desert is common to many other deserts and therefore does not deserve a proper name, just the plural *deserts*.

In a performative act of land art,[7] in 1993, Zurita carved the verses "*ni pena ni miedo*"[8] ("neither sorrow nor fear") on the desert's surface, using a bulldozer to write with, and the dry ground as the blank sheet. The bulldozer performs like a monumental amplification of Le Guin's spear and sword; the poet's stylus. In cursive font, each letter is 250 meters tall (820.25 feet), "p" and "d" are 500 meters tall (1,640.42 feet), 40 meters wide (123 feet) and 1.8 meters in depths (5.83 feet). Together, the contemporary geoglyph is 3 kilometers (1.9 miles) long and can be seen from the sky (Figure 10.2).

In the early 1990s context, in reference to his writing on the desert, Zurita affirmed[9] that at the threshold of Chile's renewed democracy, we could look at the new epoch with confidence.[10] Yet, contemplated today, Zurita's 1993 "*ni pena ni miedo*" scar on the Atacama Desert evokes a popular saying: "*sin pena ni gloria*," meaning that something went by unnoticed or is unworthy of notice. The artistic and political statement was meant to stay for ages on the surface of the land. According to Ayala Munita, Zurita inquired among geologists and he was informed that his inscription on the desert's surface would endure over 800 years (Ayala Munita 2021). But the desert has a way to demonstrate its own agency. The desert did not mind that Zurita has been profusely awarded; yet the desert did not erase all the verse. The last word in the poetic phrase: "*miedo*" (fear) remained. The question is: what is the desert trying to tell us? What does the desert fear? Why is the desert allowing this word to remain?

"Neither sorrow nor fear" is a bold statement to be spelled out in the heated desert, where living conditions are harsh on humans and the more-than-human, yet not entirely impossible. Zurita's proposal mimics the tradition of the ancestral people of the area: their immense geoglyphs, whose significance has yet to be fully understood. For some archaeological studies, they are signs in the desert to guide the first peoples' transhumance cycles.[11] For others, the stone drawings honor the

FIGURE 10.2 "Ni pena ni miedo," Raúl Zurita's land art in Atacama Desert.

Photo Guy Wenborne, 1993.

divinities on earth and sky, along the lines of what Castro calls the mirror between the material world and the spiritual world for the first peoples of the desert.

Today, 20 years after Zurita's bulldozered inscription, and with today's post-humanist sensibilities, his writing in the desert uncovers that epic impulse that Le Guin calls out. Contrariwise, the permanence of the ancient geoglyphs reveals that even though today we may have the technology to observe them from above, and even though we can master the bulldozer's strength to carve the dry land, we need a different type of wisdom and knowledge.[12] That is, the ancient geoglyphs are still there, Zurita's is continually being covered by the desert's agency (Figure 10.3).

As Ayala Munita observes, there is a tension between the permanent character that Zurita intended with his kilometric carving, but time and the materiality of his inscription on the desert's surface have had their own ways, eroding the

FIGURE 10.3 Geoglyphs of Pintados, Atacama Desert, 700 y 1500 d.C.

Photo Sznegra, 2008, Wikimedia Commons.

artistic trace. Moreover, Ayala Munita discerns that there is a parallel between Zurita's determination to write on the desert's surface about the human political condition, regardless of the desert, which is not too different from the extractivist determination to make the desert productive for the sake of anthropocentric needs. For example, *Chuquicamata* (once the largest copper open pit mine) has left its own engraving on the Atacama Desert. Of course, the magnitude of *Chuquicamata's* scar will take plenty more years to erode than Zurita's artistic action. Likewise, the abandoned saltpeter mine company towns have left their ghostly presence too. Some have been preserved as patrimonial heritage in tourist – consumption – circuits, as testimonies of a history of prosperity some will say (Garcés 2012), while others will say it's a *he-story* of extractivism; some were used as detention centers by the military after the 1973 coupe,[13] like camp *Chacabuco*, as Patricio Guzmán points out in the documentary *Nostalgia de la luz* (2010), artistically merging past, present, and future (Parham and Pippa 2016).

Professor Mauricio Ostria has written extensively about the Atacama Desert and what it takes to survive there (2014). Ostria asserts that the Chilean big northern desert – *el Norte Grande* – can be divided into three different geographic zones going from east to west: the Andes Mountain range, the desert plains or *pampa*, and the coastline. At the same time, these ecosystems construct three different human types –or rather male types–, or cultural identities, noticing that those who are not native to these lands, barely resist the challenges imposed by the desert. Yet, to become a *pampino*, says Ostria, one needs to undergo three phases: the hurting and stoic adaptation, getting lost in the desert, and being excluded from mainstream history. Alongside the diverse cultural identities related to each area, the scholar proposes that in the historical timescale, these spaces also foster different types of texts (Ostria 2009, 2018, 2020). The Atacama Desert's stories that Ostria looks at closely are stories of the humans – notably men – adapting, resisting, and surviving.

Pedro Mariño de Lobera's chronicles *Crónica del reino de Chile*[14] (1593) depicted the desert as a sterile place, with no vegetation nor water. This verdict of the desert persists in the modernization processes that advance with new settlement long after the Spanish conquest. Like Rosa Perelmuter reminds us, a few years after de Lobera, another Spanish soldier, Alonso de Ercilla, journeyed through the desert, describing it as arid and tough (Perelmuter 1998). Ercilla arrived in America in 1556, and in 1557 he arrived at what today we call Chile. He witnessed the Indigenous resistance to the Spanish conquest, which he depicted in the lengthy epic poem *La Araucana*, published in Madrid between 1569 and 1589. In the 20th *Canto*, Ercilla writes about the desert, exalting the dry tasteless matter, which he describes as so deserted and sterile that it promises endless work until the end of the days; a land that is not good at offering any liquid: "*que la seca materia desgustada [o sea, desabrida] / tan desierta y esteril que he tornado / me promete hasta el fin trabajo sumo / y es malo de sacar de un terron zumo*" (20.3) (Perelmuter 1998, 249).

Five hundred years later, these death-defying conditions are still present in the work of other authors, such as Andrés Sabella, in the mid-twentieth century. In his 1944 novel *Norte Grande*, as observed by professor Ostria, through a character who originally comes from further south, a city dweller. Sabella reflects on the social and political events of the early twentieth century in Atacama, like the workers' movement related to the saltpeter mines. Today, when the previously active company towns are unoccupied, the voices of those who recall life in these locations compose the literary canon of the desert. This is the case of Hernán Rivera Letelier, whose narratives revolve around the boom and fall of the mining settlements, but more precisely, depicting their decadence (Ostria 2018; Ostria and Heríquez 2021). Nonetheless, Mauricio Ostria laments that critics tend to focus on the biographical elements in Rivera Letelier's narratives, and more so on the picturesque components in them, rather than noticing the deeper dimension of life in the desert forwarded in Rivera Letelier's works. Yet, before Rivera Letelier, and even before Sabella, Gabriela Mistral soundly noticed how the desert forges the desert men:

> the salty desert built the Chilean man; well rooted, full achievement of American flesh. He has come out from his battle with the *caliche* crust and from his life chest to chest with the sea. When storytellers and poets want to sing about our men, they always make them, either sailor or miner, therefore revealing their natural furnace.[15]
>
> *el desierto de sal remató al hombre chileno, bien plantado, bien fundado, logro cabal de la carne Americana. Él ha salido de su pelea con la costra*

calichera y de su vida de pecho a pecho con el mar. Cuentistas y poetas, cuando quieren decir al hombre nuestro no lo hacen sino marino o minero, y dicen así sus dos forjas naturales.

(Mistral 2009, 71)

Autobiography in the Desert

Beyond the vast bibliography referred to by professor Ostria (which we have slightly discussed here), we would like to examine the contribution of three, more recent, works that coincide in using an autobiographical style yet devoting a significant portion of their narratives to telling the story of the desert. These are, in chronological order, the storied deserts by Ignacio Balcells, *La Mar: una versión de la vida ante el mar y del viaje a solas por las costas de Chile* (2001); a few years later, *Memorias del desierto* by Ariel Dorfman (2004) and then by Cristóbal Marín, *Atacama Fantasma* (2023).

Ignacio Balcells' *La Mar* has rarely been read and lacks critical reception, nonetheless, the book is an exemplary case of literature that can be vindicated today to awaken readers from environmental lethargy and aid in imagining alternative visions of our environment. Balcell's intention to put into practice a "mobile" or wandering writing-thinking process, with a nomadic spirit is key to *La Mar* (Bobadilla, unpublished). The six sections that make up the book are presented as a travel diary, a fragmentary discourse that composes, in each of its steps, a personal cartography approaching the coastal space of the Chilean mainland and island territories. The opening section, entitled "*Aprestos*" (Preparations or setouts; gathering what is needed prior to the journey), provides an outline of the itinerary that will shape the rest of the book: a journey along the coastline of the northern, central, and southern archipelagic Chile, and oceanic Chile (Rapa Nui, formerly known as Easter Island). In all these areas, the narrator explores the ways in which human beings have established relationships with the maritime boundary. This wandering through coves, beaches, and seaside villages is inaugurated with a call to movement. In the case of the first two sections, entitled "Por las costas del norte" (Along the northern coasts) and corresponding to approximately half of the book, Balcells fulfills his journey. In the first 567 pages, Balcells drives away from Santiago (our capital city), heading north towards Arica (the northernmost Chilean city). His road trip covers 2,075 km (1,290 miles) on the Panamericana Highway.[16] But his journey north towards the desert region is not a race; as he drives north, Balcells sidetracks towards every single inhabited cove. In his writing and in his lifestyle, it is evident that Balcells loved the sea and laments that Chile seems to have developed denying the sea, "*Chile de espaldas al mar*" (Chile, back to back with the sea) (Balcells 2001, 169). His wandering along the shore and offshore gave Balcells insights into the section of the Atacama Region that concludes abruptly upon the cliffs of our northern shores. The territory, the journey and his writing juxtapose, ending all at once. Balcells questions the senseless return journey that needs to trace back the same road. His unease forwards the question about what have we learnt? Could the story be told differently if the journey differed?

Sometime around the planning for the celebration of Chile's bicentennial independence (2010), Balcells read in a local newspaper that the poet Raúl Zurita was planning on another memorable land art: he would have a 22-verse poem called "*Diálogo con Chile*" written on 22 separate canvases, each one about 30 meters × 40 meters (approximately 98 feet × 131 feet), which would be fastened to the cliffs from Arica to Iquique (approximately 200 kilometers; 125 miles), facing the ocean. Balcells was annoyed. He responded by publishing a letter addressed to Zurita in the same newspaper. In his message, in a rather ironic style, Balcells manifested his concern and mocked the poet's ego. Paraphrasing Balcells' argument, he asks: "Why would it be a poetic act on the northern cliffs but infernal on the cliffs at Quintay? Because nobody lives there? That is an unacceptable justification!" (Balcells 2004).[17] Like Balcells, we do not wish to question the aesthetic quality of Zurita's verses *per se*, but the idea of such monumental intervention, the intention to leave an anthropogenic trace as such on the cliffs of a *barren area* (following Balcells' question), is horrifyingly anthropocentric, and very much "epic" in the sense that Le Guin differentiates the hero narrative from the carrier bag story. Balcells' voice not only represents the insights of someone who is aware of the environmental emergency we all face, but also a calling out to artists who might even compromise their art with the environmental crisis, yet not questioning the materiality of their own work, the means by which their messages are materialized, nor the effects and impacts of their leftovers (Casals-Hill 2023). Such blind spots reveal the contradiction of the masculine epic that Le Guin sees projected over almost all westernized cultures.

Ariel Dorfman published his book *Desert Memories: journeys through the Chilean North* in 2004. In it, using a travel diary, he plots his visits to those places that intersect with his own life and Chilean history. The same events that we mentioned at the beginning of this essay (the traces of ancestral cultures, the ghostly saltpeter mine settlements, the non-pacific war, the detention camps) are the marks that the author takes up again. From different angles he insists, by means of a *mise en abyme*, through a relationship of scales connecting time, millions of geological years; human time from the generations of the first peoples to the current settlements; and the biographical time of his own memory, his own life. This outline which articulates the book, is also understood like the general diagram of the Atacama Desert where different layers juxtapose: geographic, historical, archeological, past, and present.

Dorfman travels with his wife, they visit friends and places they had visited 40 years earlier. In these places, they recollect vestiges of their own memories. Like the other narratives highlighted in this section, the autobiographic voice is the trace that their words leave figuratively on the desert's surface, without modifying the materiality of the place. They are mindful that enough traces dwell in these places. In Dorfman's case, that voyage resolves the personal conflict, mirroring the

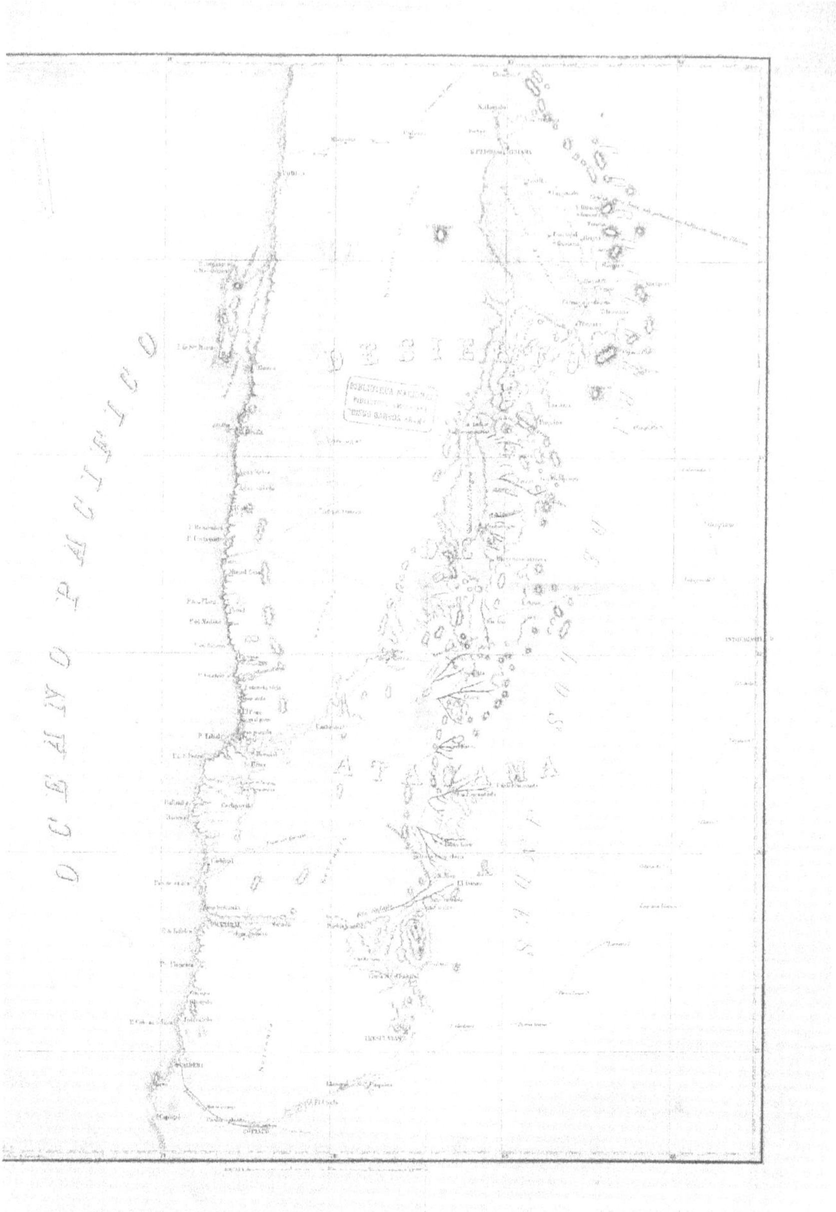

FIGURE 10.4 Atacama Desert map, Rudolph Philippi, 1850.

Photo Memoria Chilena Biblioteca Nacional de Chile, Códigos BN: MC0007220.

national tragedy. The 1973 coup had a harsh impact on these territories, not only because it hosted many detention camps, but also because even a century after the *Guerra del Pacífico*, it still is a highly militarized zone, given that it is a border zone between Perú and Bolivia. Again, Le Guin's carrier bag theory asks us: what may have been of these regions if other stories had been narrated, if other stories had been lived?

Cristobal Marín published *Atacama Fantasma* in 2023. In a style that forwards his personal story, from memories of his childhood and youth, the author traces various episodes that occurred in this vast and hostile territory over millennia. Marín's narrative spans from the *Chinchorro* cultures[18] (7,000 years ago), with their mummies, the most ancient ones that we know of so far; the voyage of the Spanish *conquistador* Diego de Almagro, the social and economic processes related to the rise and fall of the saltpeter offices, the *Guerra del Pacífico*, the voyages of the naturalists during the nineteenth century (Charles Darwin, Claudio Gay, and Rudolph Philippi), the horrific events related at different historical moments to places such as Pisagua,[19] which was employed during two Chilean dictatorships as a detention camp among other breathtaking stories. Again, these detention camps materialize the toxic masculine epic of domination; over land and over the people. The Atacama Desert is scarred by traces of a tragic fate. At present, archeologists do not only dig up for traces of ancient settlers, but also for remains of political prisoners from Pinochet's dictatorship (1973–1989), whose final resting place is still unknown (Figure 10.4).[20]

Atacama: The *He-Story*

When we ponder on the Atacama Desert as space itself, the "infinite" yet "sterile" qualities mentioned in Zurita's verses (quoted earlier in this essay) seem to be the prevalent ideas in the national imaginary. "Infinite" yet "sterile" dangerously endorses extraction. The apparent emptiness and bareness permit – or even indulge – extractivist mining, regardless of its impacts[21]; its vastness conceals atrocities like detention camps.

The same places that served as a saltpeter mine, once abandoned, became concentration camps for political prisoners during the early years of the military dictatorship (1973–1989). That is the invariable fusion of historical layers traced in every corner of this desert. For an ordinary audience, it is a matter of history, like that of many towns around the world. But for us, it is impossible to stay at that superficial semantic level; as Patricio Salinas shows in his book *Atacama, Captivity Geometry* (2020), from today's perspective, we cannot simply say *this is the life of this place*. His photographs, which are a fundamental part of his returning to that desert quest, revisiting the place where he was imprisoned for years during Augusto Pinochet's dictatorship, are a way to retain his memory of the place. Indeed, not everything has been defeated, mistreatment, and loss, but these events,

FIGURE 10.5 Atacama Desert, *Chacabuco* nitrate company's mining camp which was later used as a prison camp during Augusto Pinochet's dictatorship in the 1970s.

Photo Patricio Salinas in Atacama Captivity Geometry (2020).

political dislocation, incarceration, and exile, mark the traces over the Atacama Desert. Salinas observes the ruins, the folds left by the traces of those ways of using the space, the wound in the space itself, the irreparable fissure in the memory, looking for what the official stories have ignored or hidden in what figuratively is called "the passing of time," as if time itself could always heal (Figure 10.5).

Copper mining has left other scars on the land: like *non or post* land art, the excavation of the largest open pit mine, *Chuquicamata*, alongside its mining tailings. The "empty" desert has been terraformed by *emptying* it over 9,000 feet below surface level. Along the mine, there once was a company town by the same name, populated with over 30,000 inhabitants at its highest moment (on a surface around 42 hectares). However, due to the level of toxicity in the area, between 2003 and 2007 the population was evacuated and settled in Calama, the largest city in the area. At this point, because it made the operation cheaper given its proximity, the mining company decided to start dumping its own mining tailings on top of the former company town, erasing part of the human livelihood that had dwelled there; thus, even the *Chuquicamata* company town has been reshaped.

And we wonder, what will be lithium's inscription on the landscape to allow the flourishing of "green" energies and electric cars elsewhere look like? Today, we sense this novel extractive gaze hovering over this vast space. The future horizon of lithium exploitation is the new form of conquest, the new epic over the Atacama Desert (Bustos-Gallardo 2021). Chile has few economic alternatives: it has been copper from the nineteenth century to the present, and lithium has emerged as the new copper in the twentieth and twenty-first centuries.

Her-Story

It may seem that the Atacama Desert was mapped out only by males in their extractivist epic mode, but the truth is that the mapping of this land has been markedly masculine only in the last centuries of cultural records. In the almost imperceptible trace over millennia, the ancestral people like the *Chinchorro*, who lived and wandered upon this bioregion left traces of pastoralism, agriculture, water canalization, the remains of weavings, and other expressions of their livelihoods. In this essay, we have tried to recover the different strata that lay dormant within that immense desert that spans from the ocean to the mountains. Capturing mostly male epic stories at the literary level, but also in the history – or *he*-story – of war and domination, like Ursula K. Le Guin describes, is probably the result of our own blind spots. Nonetheless, even if the rather modern large-scale traces tell the *he*-story of the mineral exploitation of the desert (and the previous war), and the seaside industrial settlements created to export the minerals from inland, all of which were mainly populated by male workers, vanishing the footprints of millenary settlements, we mustn't ignore their heritage.

Thus, we would like to conclude bringing forth Gabriela Mistral's voice; predominantly known as a poet Nobel laureate, Mistral was also a schoolteacher in early XX century. She was born and raised at the southern edge of the desert bioregion, in the valleys that make their way from the Andes to the coast. Mistral was fully aware of the forces with which the ecozone between the desert and the valley forged her. As a public figure, Mistral raises her voice urging for the acknowledgment of other ways of telling and depicting Chilean kaleidoscopic geography. In a text written in the early 1940s, Mistral depicts the Chilean desert using colors:

> When I envision northern Chile … I see the saltpeter plains and the Atacama Desert that covers three penitent provinces, the soil has a whitewashed hazel vibration, the highland's ardent wide mirror trembles with the *puna's* roast. The desert's and the *pampa's* hazel, yellowish, and ash unify in a golden-white vibration. On the sweltering surface, there isn't any other color than this enormous fever, shining like a mirror that stalks larks; if children stare at it, they will hallucinate or even become blinded, just like the morning bird. The sun burnt everything; waters weren't enough to fight back. The ancient subtropical forest that they say covered the desert was beaten by aridity and water hid among gravel and pebbles, where it ended up lost.[22]
>
> *Cuando veo el norte de Chile ... veo... la pampa salitrera y el desierto de Atacama que cubren tres provincias penitentes, la tierra tiene un pardo*

> *blanquecino de reverberación, un ardor de espejo ancho que en las partes altas tiembla un poco con el sollamo de la puna. El pardo, el amarillento y el ceniza de la pampa y del desierto se unifican en un blanco dorado de reverberación. En el suelo que arde no hay más color que el de esta larga calentura; brilla como el espejo que caza las alondras y si los niños lo pueden mirar fijamente se alucinarán o cegarán lo mismo que el pájaro cantor. El sol quemo todo allí donde las aguas no eran tantas como para luchar con él. El antiguo bosque subtropical que dicen que cubrió el desierto fue vencido por la aridez y el agua se escondió en los cascajales y acabó perdiéndose en ellos.*
>
> (Mistral 2013, 89)

Mistral is aware of that millenary layer where the bioregion was once green. She is also fully aware of the hardships of desert life, yet she depicts a vibrant landscape; neither a void, nor merely a mineral supply, and definitely, not a stage for war.

In the poetry collection *Poema de Chile* (1967) in the poems called "*Desierto*" (Desert) and "*Canción de cuna del ciervo*" (Deer's lullaby), Mistral refers to her partners in the journey through Chile, an Indigenous boy and a Chilean deer, a *huemul*, as "*tacneños*," associating them to the city of Tacna. Tacna is the southernmost city in Perú, once the political borders were reconfigured as a result of the *Guerra del Pacífico* which we have mentioned earlier in this essay. For those acquainted with Mistral, it is puzzling that being a geography teacher, when she refers to these beings, Mistral dismisses the geo-political boundaries; while singing the "long poem of Chile," what counts for her is the bioregion, regardless of their geopolitical limits (Casals 2017). While Mistral sees a bioregion where humans and the more-than-human coexist, therefore writing a caring story of the desert, for the large majority, our desert is a place of male epics, extractive operations, and war confrontations. Would this desert tell a different story if a female majority had populated it and if those female voices had told its story?

In 2003 Ursula K. Le Guin published an English anthology of Gabriela Mistral's poems which Le Guin selected and translated over many years. Among her own reflections regarding Mistral, Le Guin ponders:

> For the last several years I have been working on a translation of the Chilean poet and Nobel Prize winner Gabriela Mistral. (She was about a half-generation older than Pablo Neruda, and I think he learned a great deal from her.) Her Nobel was the first ever given to a South American writer. … I fell in love with Mistral when I first read her, and the love only grows deeper as I work with her words and try to understand and to reflect her passion and her art.
>
> (Le Guin 2003)

Politely, yet assertively, Le Guin positions Mistral as an influence on her and for our second Nobel poet. It is Mistral's storytelling that Le Guin seeks to comprehend.

Looking in a different direction, in Ursula K. Le Guin's short story "Sur" (in *The Compass Rose*, 2015), the speculative fiction proposes a group of women set to explore and reach the Southern Pole, crossing that other desert. Their journey is secretly planned, and the story is well informed by the dates and names of the ships and crew that aided the expeditions of non-fiction male legendary explorers (some who made it to the South Pole, some who did not). When the fictional women reach their destination, being the first modern humans to make it there, they celebrate in circular dances over silvery-white ice. But as they leave the epic center of the largest, driest desert on Earth, they decide to leave no trace, adopting an ethics of care for the immense ice continent itself, as well as for the white males who were yet to come and proclaim themselves as the first men to reach the Pole in the turn of the century's exploration and conquest competition. Understood from the paradigms of the *he*-stories of explorers and conquerors, Le Guin's tale depicts a voyage that goes by intentionally unnoticed (*sin pena ni gloria*).

The stories we have told of the Atacama Desert include a non-pacific expansionist war, company towns, ghost towns, mining terraforming, political detention centers, and even artistically minded interventions that have left a variety of scars on the desert, each one telling their own epic. We wonder if we had a similar outlook as that of Le Guin and Mistral, we may imagine what would have become of the Atacama Desert if the layout of *her*-story had been told recollecting seeds and roots. It is likely that even the economic history of Chile would be different. It is not about being naive, but we could also look differently at the storied facts and elements intertwining with Atacama.

Notes

1 This article presents a collaborative finding from a research project funded by the grant Fondecyt Regular nº1230624 from Pablo Chiuminatto "Another end of the world is possible"; and Andrea Casals-Hill's research "How Radical Can Hope Be" Fondecyt Iniciación 1200236, ANID, Chile (National Research and Development Agency, Chile).

2 Our translation. Original in Spanish: *"... el imaginario colectivo responde ... a la imagen de un lugar cálido y seco, azotado por un ardiente sol durante el día, y dominado por un desapacible viento y la caída de las temperaturas durante la noche"* (Rodríguez 2023).

3 Our translation.

4 *Purgatorio* is a poetry collection by Raul Zurita, who is possibly the most important currently living Chilean poet. *Purgatorio* (1979) is the first part of a poetic trilogy, followed by *Anteparaíso* (1982), 14 years later. In 1993 he concluded the trilogy with *La vida nueva* (1993).

5 Our translation.

6 Zurita, 30.

7 In 1979 Zurita brought together an artists' collective named CADA (for its acronym in Spanish: *Colectivo de Acciones de Arte*). CADA made large scale, monumental public art as a form of resistance against Pinochet's dictatorship.

8 "*Ni pena ni miedo*" are the last verses in the third book of the trilogy, *La vida nueva*.

9 That is how he is quoted by the local journal *La Estrella de Valparaíso* in 1993, in the brief note titled "Mensaje escrito en el desierto" (http://www.bibliotecanacionaldigital.gob.cl/colecciones/BND/00/RC/RC0003351.pdf).

10 Zurita's optimism can be traced in the titles of the poetry trilogy mentioned earlier: *Purgatorio* (1978) and *Anteparaíso* (1982) were published during the dictatorship, while *La vide nueva* (1993), "The New Life", was published three years after democracy was regained in Chile.

11 Transhumance refers to the cycles of moving livestock from a grazing ground to another; usually following summer to winter cycles, but also to allow the pastures to recover.
12 The Chilean artist Cecilia Vicuña calls learning from ancient cultures a future-making capacity or habit (*una capacidad futurante*) which enables us to look at the future more wisely.
13 On September 11 in 1973 the Chilean military overthrew the democratically elected socialist government led by Salvador Allende, imposing 17 years of dictatorship by Augusto Pinochet. In 1991, a year after democracy was regained, the Retting Report affirmed that over 2,000 people were detained for political reasons; many were killed. To date, half are still "disappeared". Later, the Valech Report, published in 2004, received almost 36,000 testimonies; over 27,000 persons were regarded as direct victims, most of them had been tortured.
14 The Chronicles were first published 1865 in a larger history collection, but the manuscript was written by mid to late XVI century.
15 Our translation
16 The Panamericana Highway is the main highway in Chile. Given the long and narrow geography of the country, it is impossible to travel from north to south, and back again, avoiding it.
17 "¿Por qué va a ser poético en el acantilado nortino lo que en el acantilado de Quintay sería a todas luces infernal? ¿Porque allá no vive casi nadie? ¡Inaceptable justificación!" (Balcells, "Carta abierta a Raúl Zurita 2004).
18 The Chinchorro people were a fishing culture that lived in the coastal region of what is now Perú and Chile, between the years 9,100 to 3,500 years BP (7,000 to 1,500 BCE). See: Museo de Arte Precolombino, Chinchorro (https://precolombino.cl/wp/en/archivo/coleccion/area/andes-sur/chinchorro/)
19 Pisagua is a coastal town in northern Chile. In 1879 it was occupied by Chilean troops during the *Guerra del Pacífico*. During the saltpeter boom, Pisagua became a major port. After its decline, because of its isolation, between the desert and the sea, Pisagua was used as a detention camp; most notably, by Pinochet's dictatorship.
20 In Chile they are "known" as the *detenidos-desaparecidos*; that is, "detained and disappeared". Like the Mothers from Plaza de Mayo in Argentina, female relatives of the detenidos-desaparecidos still comb the desert searching for their lost ones. The British pop musician Sting wrote "They dance alone" (1987) in homage to the women who dance alone the Chilean national dance (*cueca*) with a photograph of their beloved *detenido-desaparecido*.
21 Now-a-days the apparent emptiness and bareness permit –or even indulge– dumping fast fashion trash from overseas surplus production in Atacama regardless of its impacts. And the national imaginary still believes there is no impact because the desert is empty.
22 Our translation.

References

Ayala Munita, Matías. 2021. "'Ni pena ni miedo', de Raúl Zurita: desierto y erosión." *Estudios de Teoría Literaria-Revista digital: artes, letras y humanidades*, 10.21, 103–116.

Azua-Bustos, A. et al. 2022. "The Atacama Desert in Northern Chile as an Analog Model of Mars." *Frontiers in Astronomy and Space Sciences*, 8, https://doi.org/10.3389/fspas.2021.810426.

Balcells, Ignacio. 2001. *La Mar*. Andrés Bello.

———. 2004. *Carta abierta a Raúl Zurita*. https://www.escritores.cl/base.php?f1=varios/carta_zurita.htm

Barnum, Dennis W. 2003. "Some History of Nitrates." *Journal of Chemical Education*, 80.12, 1393.

Bobadilla, Rodrigo. 2023. "La estela de un nomadismo oceánico: Notas para navegar La mar de Ignacio Balcells", in the PhD thesis unpublished. Pontificia Universidad Católica de Chile.

Bustos-Gallardo, B. et al. 2021. "Harvesting Lithium: Water, Brine and the Industrial Dynamics of Production in the Salar de Atacama." *Geoforum*, 119, 177–189.

Casals-Hill, Andrea. 2017. "La loca ecología de Gabriela Mistral." *Taller de Letras*, N. 60, 9–17.

———. 2023. "Proyecto 'Arte en riesgo': una experiencia interdisciplinar." *Humanidades: Revista de la Universidad de Montevideo*, No. 13, 39–62.

Castro, Victoria. 2022. "Vida en el del desierto de Atacama, una visión desde el presente." In Ch. Signorio y L. Ortúzar, *Diálogo entre desiertos: Festival de Ciencia Puerto de Ideas Antofagasta 2015–2019*. Puerto de Ideas, 93–103.

Dorfman, Ariel. 2004. *Desert Memories: Journeys through the Chilean North*. National Geographics.

Doyle, S. 2021. "Atacama Large Millimetre Array (ALMA)[Engineering Places-IET@ 150]." *Engineering & Technology*, 16.10, 62–63.

Garcés, E. 2012. *Las ciudades del salitre*. Santiago: Orígenes. And see, "Asentamientos de la minería y desarrollo territorial" en Alonso, P., *Deserta. Ecología e industria en el des erto de Atacama.* Santiago de Chile: Ediciones ARQ, 104–111.

Garreaud, René D. et al. 2010. "Andean Uplift, Ocean Cooling and Atacama Hyperaridity: A Climate Modeling Perspective." *Earth and Planetary Science Letters*, 292.1–2, 39–50.

Gayo, Eugenia M., et al. 2019. "Geohistorical Records of the Anthropocene in Chile." *Elementa: Science of the Anthropocene*, 7, 15.

Le Guin, Ursula K. 2003. *Select Poems from Gabriela Mistral*. Retrieved August 28, 2023. https://www.ursulakleguin.com/selected-poems-of-gabriela-mistral

———. 2019. *The Carrier Bag Theory of Fiction*. Ignota Books, 2019.

———. 2015. *The Compass Rose*. Hachette UK.

Lobera, Pedro M. de. 1865. *Crónica del reino de Chile, escrita por el capitán Don Pedro Mariño de Lovera. Dirijida al Exmo. Señor Don Garcia Hurtado de Mendoza... Reducida nuevo método, y estilo por el padre Bartolomé de Escobar* (Vol. 6). Imprenta del Ferrocarril.

Marín, Cristóbal. 2023. *Atacama fantasma*. Debate.

Mistral, Gabriela. 1967. *Poema de Chile.* Editorial Pomaire.

———. 2009. "Pequeño mapa audible de Chile (1931)." In *Chile País de Contrastes*. Biblioteca Fundamentos de la Construcción de Chile.

———. 2013. "Mapa coloreado de Chile." In *Caminando se siembra. Prosa inédita*. Ed. Luis Vargas Saavedra. Lumen.

Montt, I., et al. 2023. "Chinchorro Fibre Management in the Atacama Desert and Its Significance for Understanding Andean Textilization Processes." *Journal of Anthropological Archaeology*, 71, 101530.

Núñez, Lautaro, et al. 2010. "Sequential Analysis of Human Occupation Patterns and Resource Use in the Atacama Desert." *Chungara, Revista de Antropología chilena*, 42.2, 363–391.

Ostria, Mauricio. 2009. "La identidad pampina en Rivera Letelier." *Acta literaria*, 30, 67–79.

———. 2018. "Desierto y Literatura en el Norte Grande: una mirada ecocrítica." *Anales de literature chilena*, No. 30.

———. 2020. "El desierto en la literatura chilena. Una mirada ecocrítica." *Taller de Letras*, 67, 197–215.

Parham, J. and Marland, P. 2016. "Every Night the Universe Passes over Santiago: Transational 'Ecocinema' and Visual Environmental Histories in Patricio Guzmán's *Nostalgia for light.*" *English Studies in Latin America*, No. 12.

Perelmuter, Rosa. 1998. "El desierto en La Araucana." *Calíope: Journal of the Society for Renaissance and Baroque Hispanic Poetry*, 4, 248–257.

Reyes, Salvador. 1969. *Andanzas por el desierto de Atacama*. Zig-Zag.

Rodriguez, Hector. 2023. "El desierto más grande del planeta es un mundo de hielo." https://www.nationalgeographic.com.es/medio-ambiente/desierto-mas-grande-planeta-esmundo-hielo_19441

Sabella, A. 1944. *Norte Grande*. Orbe.

Sagredo, Rafael. 2014. "Un paisaje inquietante." In Peliowski, A., and Valdés, C. *Una geografía imaginada. Diez ensayos sobre arte y naturaleza*. Ediciones Universidad Alberto Hurtado-Metales Pesados.

Salinas, Patricio. 2020. *Atacama: Captivity Geometry*. Saposcat.

Santoro, Calogero M., et al. 2017. "Continuities and Discontinuities in the Socio-Environmental Systems of the Atacama Desert during the Last 13,000 Years." *Journal of Anthropological Archaeology*, 46, 28–39.

Standen, V. G., and Arriaza, B. T. 2014. "The Chinchorro Culture: Hunter, Gatherers and Fishers of the Atacama Desert Coast." In *The Chinchorro Culture: A Comparative Perspective. The Archaeology of the Earliest Human Mummification*. UNESCO, 35–52.

Valech Report. 2004. Comisión Nacional sobre Prisión Política y Tortura. In *Informe de la Comisión Nacional sobre Prisión Política y Tortura (Valech I)*. Salesianos.

Workman, T. Race, et al. 2020. "Landscape Evolution and the Environmental Context of Human Occupation of the Southern Pampa del Tamarugal, Atacama Desert, Chile." *Quaternary Science Reviews*, 243, 106502.

Zurita, Raúl. 1979. *Purgatorio*. Universitaria.

———. 1982. *Anteparaíso*. Editores Asociados.

———. 1993. *La vida nueva*. Universitaria.

11

FROM HANDBACK TO LANDBACK

Lessons from Uluṟu

Tom Lynch

The five-hour drive from Alice Springs to Uluṟu passes through an impressive variety of desert terrain, increasingly arid as one heads south on the Stuart Highway and then west on the Lasseter. When, after four hours, the summit of Mount Conner rears on the horizon, most visitors mistake it for Uluṟu.[1] Unlike the rounded outline of Uluṟu, however, Conner is a classic mesa, flat topped with steep, nearly sheer sides, jutting above an utterly flat horizon. In the American Southwest it would certainly be called Conner Mesa, but the term "mesa," being a legacy of Spanish settlement in the North American deserts, is not used in Australia.

I pull off the road for a break at the Mount Conner lookout, a popular viewpoint. From this location, one gets a real sense of the landscape's aridity. Rich red sands, sprinkled with tussocks of spinifex grass, gray-green acacia shrubs, and short mulga trees, stretch off toward the distant Conner. Behind me gleams blindingly white salt flats, the eastern edge of the vast Lake Amadeus basin. Simmering beneath a cloudless blue, it's a mirage landscape. The terrain is beautiful but daunting. This is a region that bedeviled European explorers and later prospectors. In 1872 Ernest Giles was stopped in his tracks by the crust of the salt flats, which could not bear the weight of his horses. This highway itself is named for Lewis Lasseter, who perished in 1931 from starvation while searching for a legendary lost gold reef, a warning, perhaps, to those who would underestimate the perils of this lovely landscape. Don't wander off.

After another hour of driving, with Xavier Rudd blaring from the speakers of my rented Toyota, I reach the park. I am greeted by a sign adorned with the distinctive and iconic artwork of the Central Desert Aboriginal communities:

Welcome to Aboriginal Land: Uluṟu-Kata Tjuṯa National Park.

DOI: 10.4324/9781003397212-15

This is followed by the welcome repeated in the two related languages of the Anangu people of this region:

> *Pukul ngalya yanama Ananguku ngurakutu* (Yankunytjatjara)
> *Pukulpa pitjama Ananguku ngurakutu* (Pitjantjatjara)

And at the bottom of the sign:

> *This is Anangu land and you are welcome. Look around and learn in order to understand Anangu and also understand that our culture is strong and alive. – Traditional owner*

One might be forgiven for being a bit cynical about this. Cynicism is too frequently warranted in cases of what might seem to be tokenistic gestures made by settler states to the Indigenous peoples they have colonized. In this case, however, we might want to temper our cynicism a bit.

In a 2021 article in *The Atlantic*, "Return the National Parks to the Tribes," David Treuer (Leech Lake Ojibwe) makes the case that the United States should return ownership of all US National Parks to the original Indigenous owners who had been displaced during the creation of the parks. He argues that all of the national parks "were founded on land that was once ours, and many were created only after we were removed, forcibly, sometimes by an invading army and other times following a treaty we'd signed under duress." Because of this history of land theft, Treuer concludes that "for Native Americans, there can be no better remedy for the theft of land than land. And for us, no lands are as spiritually significant as the national parks. They should be returned to us. Indians should tend—and protect and preserve—these favored gardens again."[2]

In the United States, even if we are sympathetic to Treuer's proposal, we might nevertheless consider the return of national parks to Native ownership to be implausible. It's hard to imagine entering a US National Park and passing a welcome sign like the one that greeted me at Uluru-Kata Tjuta. But the return of park lands to Indigenous control has been happening for many decades in other settler societies, Australia being the most significant example. Treuer himself makes brief mention of the fact that Australia (as well as New Zealand) have returned many of their national parks and other protected public lands to Indigenous ownership, and he cites Uluru-Kata Tjuta National Park (formerly Ayers Rock-Mount Olga) as a notable example. But he missed an opportunity to provide details about how this handback of Uluru occurred, of how the Aboriginal ownership and management of the park works, of how such change alters a visitor's experience of the place, and of how this return of land has affected the local Aboriginal community. The circumstances regarding the relationship between national parks and Indigenous people in the United States and Australia are by no means identical; nevertheless, they are similar enough that the Australian program of land return and joint park management offers, if not a specific blueprint, at least inspiration and guidance for similar efforts in the United States. The main obstacle, I suspect, is a general ignorance among Americans, even cosmopolitan academics and globally aware activists, of things Australian.

It is now well established in the field of settler-colonial studies that national parks, monuments, and similar protected areas are distinctive forms of spatial demarcation of land that derive from the settler-colonial imaginary. The earliest national parks were all established in settler societies: America's Yellowstone in 1872, Australia's Royal in 1879, Canada's Banff in 1885, New Zealand's Tongariro in 1887, and South Africa's Kruger—as the Sabi Game Reserve—in 1898. All these cases resulted in a loss of Indigenous sovereignty, imposed at least some degree of restrictions on Indigenous peoples' access to and use of the land, and in many cases resulted in the forced removal and banishment of their entire communities. And yet many of the lands of most vital importance for environmental protection and wildlife restoration in the United States and Australia (as well as in other settler societies) are these very same protected public lands. This leaves environmentalists with a quandary. On the one hand, these lands are key elements in a network of environmentally important landscapes, and one would be loath to remove such protections and subject these beloved and ecologically significant places to an uncertain future. On the other hand, the fact that such lands were typically stolen from Indigenous peoples means that their existence rests on a historical—and very much continuing—injustice that many settlers feel compelled to mitigate and, when possible, redress.[3] In the United States little has been done toward decolonizing parks, monuments, nature refuges, and similar protected lands. Indeed it is probably no coincidence that when Indigenous people have been most pro-active in promoting the development of a protected area—the Bears Ears National Monument—their efforts provoked such strident political backlash that for a time the monument's borders were reduced by 85 percent. This reactionary backlash suggests that it was not simply the creation of the national monument that opponents found objectionable but also the assertion of Indigenous sovereignty that the creation of the monument gestured toward.[4]

Because such federal parks and monuments are usually of great ecological value, highly scenic, and, for many people—both Indigenous and settler—numinous if not indeed sacred places, they represent powerful and visible locations at which to attempt reconciliation activities such as landback efforts.[5] As federal public lands, their return to Indigenous control would not conflict with private property rights nor state-level sovereignty interests (though the usual people would no doubt complain).[6] The high visibility of such places, both as national icons and as tourist destinations—some of which see several million visitors a year—can serve to educate the general public about the possibilities to redress settler-colonial land theft, raising consciousness and laying the sociocultural groundwork for other landback and redress efforts. Landback depends upon whose stories get told. Whose stories count. Whose stories underwrite our relationships with land. Whose stories dictate ethical and spiritual relations to place. Stories, that is, serve as evidence to sanction landback. And the stories we tell of landback efforts are themselves vital place-based tales that can serve to inspire further landback projects.

The dramatic rock formation referred to as Uluru by the local Aboriginal community, the Anangu, first entered the stories of Europeans in 1873, when the surveyor William Gosse christened it Ayers Rock, in honor of Sir Henry Ayers, who had served as premier of South Australia. In the ensuing decades, only a few white Australians visited Uluru, mainly scientific explorers such as the 1889 party led by William Henry Tietkens, the first person to photograph the rock, and the 1894 Horn Expedition, led by the anthropologist Baldwin Spencer. Slowly but incessantly, however, Australian settler-colonialism crept into the region, primarily in the form of Christian missions, pastoral enterprises, and gold prospecting, which became the basis of the predominant stories we settlers have told ourselves about the land. In 1920 the desert lands in the Northern Territory around Uluru were officially designated as the Petermann Aboriginal Reserve, part of a larger Central Australian Aboriginal Reserve that included adjacent lands in the states of Western and South Australia. However, when tourism to "the rock" increased following the development of primitive road access in 1948, Uluru and the surrounding land were transformed into Ayers Rock National Park in 1950. In 1958, the rock and adjacent areas were officially excised from the Aboriginal reserve and Ayers Rock-Mt. Olga National Park was established.[7] In the 1960s and 1970s, tourism began to boom, with improved roads, a series of trails, a handrail to assist in the climb to the top of the rock, a motel, campgrounds, and an airstrip. (It was during this chaotic period that the infant Azaria Chamberlain was killed by a dingo, probably one that had been fed by visitors.) Tourism operators even began to pressure the government to remove the Anangu people from the park, and the government took steps both to dissuade them from remaining and to entice them to settle at other locations beyond the park boundary, such as the Docker River settlement.

At the same time that the government was slowly removing Anangu people from the park, however, Aboriginal people throughout Australia were organizing a countervailing land rights movement. In 1976 the Northern Territory Aboriginal Land Rights Act was passed, enabling Indigenous communities in the Territory to file Native Title claims on their traditional homelands.[8] Devolving from this and related acts, several court cases and land claim adjudications followed. The outcomes in these cases often hinged on the collection of Aboriginal stories linking the members of the community to the places being considered under the title claims. The stories acted as a sort of legal documentation demonstrating people's enduring connections to place. This process culminated in October 1985 when Australia's Governor-General, Sir Ninian Stephen, officially handed the title deeds for the park to representatives of the Anangu Traditional Owners.[9] A major celebration was held, and the event became widely and iconically known as the "handback."[10] As part of the deal for gaining title, the Anangu then leased the land to the federal government's Director of National Parks for a period of 99 years to continue to be managed as a park. This arrangement created a ten-member management board, including six Traditional Owners chosen from among the local community, three members of the federal government, and one ecologist. The Aṉangu continued to reside in the park, near the rock but shielded from tourist intrusions in a community called Mutitjulu. And the agreement allowed traditional hunting and gathering practices to continue. The Aṉangu also restored the ancient cultural practice of burning to diversify habitat and to reduce the danger of uncontrollable wildfire.[11]

Of course, the fact that the handback was conditional on leasing the land back to the federal government to run as a park indicates that the Aṉangu did not gain full sovereignty; nevertheless, they maintained majority control of the management board. The Uluṟu handback was a momentous event.[12] Since the return of Uluṟu to Aboriginal ownership the Australian federal government, as well as various state governments, have made similar arrangements with many of the Traditional Owners upon whose lands their parks were established. Today many Australian parks and other natural areas are owned by the local Aboriginal communities and jointly managed with the federal or state governments. And inspired by these examples, Australia has established a new form of land tenure, the Indigenous Protected Area. These areas are roughly akin to Indigenously managed wildlife refuges where traditional ecological knowledge and spiritual traditions are combined with scientific management principles.[13] Because of the overall success of such efforts, new parks and protected areas continue to be added to these various joint management programs.[14]

No landscape feature in Australia serves as a more iconic representation of the nation than Uluṟu, and so its handback has had enduring repercussions, some highly visible, such as the name change and the welcome sign I passed, and some involving a less obvious but perhaps more important psychic reorientation to ideas of place and belonging among both the Indigenous communities and settler Australians, and even among many of the international tourists, such as myself, who visit Uluṟu. The experience visitors have of this desert landscape is now largely filtered through the lens provided by the Aṉangu. By creating a system of signage, ranger guided tours, pamphlets, and related park infrastructure the Aṉangu tell visitors what we should notice, explain what they deem to be important, point out what places we should respectfully avoid, and provide ways to interpret what we experience. They tell us whose stories count in this place. The welcome sign that greeted me serves not only as a friendly welcome but also as a claim to authority, knowledge, and control.

<<<>>>

Kaarnka, the raven, calls loudly as I walk across the parking lot to the trail head, echoing off the massive rock face of Uluṟu that looms nearby. The voice of the Australian raven (*Corvus coronoides*) has weirdly human tones and inflexions, providing an uncanny welcome to what many find to be an uncanny place. It's not as crowded here as I expected, an unanticipated blessing. I've lucked out coming in

May, before the peak tourist season and between the major school holidays in April and July during which Australian families flock to their country's parks.

Near the start of the trail that encircles Uluṟu, a spur route heads up onto the massive rock, leading all the way to the summit. To get to that route visitors must pass a sign:

> **PLEASE DON'T CLIMB.**
>
> We, the Aṉangu traditional owners, have this to say
> Uluṟu is sacred in our culture. It is a place of great knowledge.
> Under our traditional law climbing is not permitted.
> This is our home.

A more detailed explanation continues further below. This sign is worded as a request, so people are allowed to ignore it. And many do.[15] As I pause to read the sign myself, I notice a woman standing beyond it, at the base of the rock, looking up, excitedly waving at her husband and child 100 yards away along the summit trail. The father and child cheerfully wave back. Beyond them the trail extends as a gray line crossing the reddish-orange of the massive rock, the rusty patina of oxidation having been worn away under the vibram soles, sneaker prints, and sandal tracks of so many visitors. The ease of the climb is enhanced by the handrail that was installed in 1966, and I spot several people farther up the trail, clinging to the rail, growing increasingly tiny as they near the summit. The slightly derogatory Aṉangu term for tourists, especially those who climb the rock, is *minga*, meaning ant. And from this perspective, the term seems very apt.

Standing here I recall a family drama from 15 years ago, the first time I had visited Uluṟu, with my wife and our two boys, then ages 5 and 8.

> "Why can't we climb? We want to climb it."
> "Well, because the Aboriginal people who own it ask us not to climb."
> "But look, look at all those other people climbing!"
> "Yes, well they aren't being respectful."
> "Puhleeze, I bet it's cool up there."
> "I'm sure it is."
> "I bet there's a great view."
> "Probably a very good one."
> "No fair, no fair."

And so my wife and I tried to reason with our two boys about why we, unlike seemingly everyone else, would not be climbing this super cool trail to see a spectacular view from the top of the rock but instead would be making the long boring hike around it. Not the best way to begin a hike with small children, alas. Today I'm alone, so I get no grief as I turn away from the temptation of the summit trail and head out on the Base Walk, a ten-and-a-half-kilometer circuit.

The first section of this trail is designated as the Mala Walk. One of the principle *Tjukurpa* or Dreamtime stories of Uluru involves the story of Mala, the rufous hare-wallaby (*Lagorchestes hirsutus*), a small rabbit-sized marsupial.[16] Mala were once an abundant species throughout the Australian Central Deserts, but ecologists now consider them to be extinct in the wild. The population declined primarily due to predation by introduced cats and foxes as well as the loss of habitat with the reduction of traditional Aboriginal burning practices. The Uluru-Kata Tjuta National Park now operates a reintroduction program, seeking to build up a protected population before eventual release. The powerful spiritual presence of Mala juxtaposed against the absence of the physical animal lends this place a tender poignancy.

The Mala Walk is a short and popular section of the much longer Base Walk. Most people stroll out to Kantju Gorge and then return to the parking lot, a 2-kilometer trek. I should have arrived early enough to take the Aboriginal ranger-guided tour along this stretch of the trail, but, sleeping in after a long day, didn't make it. Instead, as I stroll along this trail alone, later in the morning, I consult the brochure I had picked up at the visitor's center, "An Insight into Uluru," which turns out to be a more important document than I had realized at the time. This brochure includes some of the Anangu stories of the place and directs my attention to key locations:

> In the beginning, Mala men, women and children must travel a long way from the west and north to reach Uluru. When they arrive they camp at separate sites from one another in groups of young men; old men; young and single women; and old and married women. They do this because they are here for an Inma (religious ceremony).
>
> Some Mala men, who come from the west, carry the ceremonial pole Ngaltawata. They scramble quickly to the top of Uluru and plant the pole in the ground at the most northern corner to begin the Inma. From this moment on, everything becomes a part of the ceremony. Even everyday jobs like: hunting; gathering and preparing food; collecting water; talking to people; or just waiting, are now done in a proper way for ceremony. This has become Law for men, women and children ever since.[17]

The route that the Mala men followed to the summit in this tale is the same route that visitors climb today. (Really, I imagine it's the only route by which one *could* walk to the summit.) Only those individuals who have inherited the rights to and been formally initiated into the Mala Tjukurpa are allowed to walk this route, hence the request for others not to climb.

As the brochure explains, a conflict broke out when the Mala were interrupted in their ceremony by a neighboring group, the Wintalka. This group asked the Mala to join their ceremony, but, having already begun their own, the Mala refused. Insulted, the Wintalka unleashed Kurpany, "an evil, black dog-like creature," to attack the Mala and chase them away. The trail I am strolling passes several of the key sites in this story, such as the Kitchen Cave where the Mala prepared their meals. These rock shelters—formed by wind erosion—are adorned with layer upon layer of pictographs, more recent ones obscuring older ones. The rock shelters are all shady and cooler than the surrounding landscape and would undoubtedly have been pleasant places to dwell during summer heat as well as providing adequate shelter during rainy times.

Interestingly, there is a story about the creation of this brochure of stories I am carrying. In his book *The Rock: Travelling to Uluru* (1994), Barry Hill describes how during his time researching at Uluṟu he was guided by Tony Tjamiwa, one of the A̲nangu Traditional Owners. Hill explains how Tjamiwa had told anthropologist Susan Woenne-Green that

> The booklet exists because people like Tjamiwa got fed up with tourists who keep coming and 'never see anything.' … All around there were rocks, trees and other things to do with the *mala* story. These marks, these tracks is where they *really* were, Tjamiwa had said in one of the many exchanges that went into the production of the booklet.

Hill explains how

> the booklet went through several drafts of consultation with the designer as well as owners of the stories. The task was to place the story properly in the context of the walk itself; to position the narrative in the landscape in a way that oriented people towards the story in a way that fitted Anangu's.[18]

If my experience is any example, the brochure has indeed properly placed the story, although I realize the short narrative summaries in the brochure are only a tiny dimension of a much larger, richer, ritual epic and ceremonial network.

In addition to the tale of the Mala, many other stories converge here as well, including the tale told by science. Looking at the imposing size and symmetry of Uluṟu from a distance, and then standing close and reaching to touch its rough red surface while craning one's neck to peer straight up, one naturally wonders how such an unlikely place came to be. The story that science tells is that the feldspar-rich arkose sedimentary rocks that formed Uluṟu were laid down and compressed beneath a shallow sea in the Amadeus basin about 550 million years ago. Through a series of faulting and erosion episodes, the rock (technically an inselberg) began to emerge as a single unit above the surrounding, eroding landscape about 300 million years ago. Dinosaurs would have walked around it. Aboriginal people moved into the region about 30,000 years ago, give or take, and Uluṟu has likely been an important location for them ever since. The story these people themselves tell is that, in the Tjukurpa, their ancestors created the rock. According to the rendition published by Robert Layton, "Uluru (Ayers Rock itself) was built up during the creation period by the two boys who played in the mud after rain."[19] Later, different portions of the rock, then still pliable, were shaped into their present form by different ancestral beings who had their own adventures in this landscape. This northern section, where the trail begins, was shaped by the adventures of the Mala and the landscape around one teems, if not with the actual animal, alas, at least with Mala stories and pictographs. In my stroll, I pass several other rock shelters. Most striking is Kulpi Watiku, the senior men's cave, where the Mala men prepared themselves for the Inma. Like the Kitchen Cave, it too is adorned with multiple layers of pictographs.

The Mala Walk ends at the Kantju waterhole, a splash pool at the base of a tall, narrow, trickling waterfall. The brochure tells me that "the wind whispers through the trees here. This has always been a quiet area. When A<u>n</u>angu walk along this path to collect water from Kantju, we always move silently. Very close are sites of great spiritual importance to A<u>n</u>angu men." It is a lovely spot, and the wind is indeed whispering through the trees. This is, I agree, a good place to be quiet, and I'm grateful when the two chatty French tourists depart. Various birds sing in the trees. Trickling water splashes softly. The contrast of lithe green eucalyptus trees against a background of the massive, brilliantly red rock is stunning. Wildflowers bloom all around. This is one of the few reliable water holes in a vast territory, and it feels like an Edenic oasis. Every once in a while, one experiences a place where all the features of life on planet Earth seem to be gathered into an especially comforting, nurturing, congenial location. In such a spot all the elements are arrayed for our delight: water, food, shelter, beauty, mystery. And one thinks, yes, this is it. Yes, this is what our lives on planet Earth are supposed to consist of. Today, at least, Kantju feels like such a place.

The vegetation growing around Uluṟu is almost as impressive as the rock itself. Some of the plants are conveniently labeled with signs, saving me the trouble of trying to find them in my field guides. The signs give preference to the A<u>n</u>angu name, then provide the common settler name, and finally the scientific designation:

- kuniya-kuniya, broom ballart, *Exocarpos sparteus*
- muur-muurpa, desert bloodwood, *Eucalyptus terminalis*
- i<u>t</u>aṟa, river red gum, *Eucalyptus camaldulensis*
- altarpa, blue mallee, *Eucalyptus gamophylla*
- i<u>l</u>i, rock fig, *Ficus platypoda*
- arngu<u>l</u>i, bush plum, *Santalum lanceolatum*

As these names suggest, many of these plants offer fruits that make for good bush tucker.

With all the vegetation, reliable water, and abundant fruits, this place is an oasis not just for people but for birds as well. The park's birdwatching flyer says 178 species call Uluṟu home at least some of the time. I spot a few: *warutjilyarpa* (grey-headed honey-eater), *tjalpu-tjalpu* (black-faced woodswallow), *miititi* (crimson chat), *patupiri* (fairy martin), and *mirilyirilyi* (splendid fairywren). As one might be able to tell, most of the Aṉangu names are onomatopoetic, imitating the sound the bird makes. One bird's call is particularly distinctive, a gentle chime, repeated over and over, and echoing around the rocks and through the trees, a lovely effect but making the bird itself hard to find. I'm not surprised to read that this bird, the *panpanpalala*, or crested bellbird, plays a role in the Mala story, functioning as a messenger. Its song does have the rhythm of its name: pan-pan-pa-la-la, pan-pan-pa-la-la. And the English name "bellbird" is descriptive as well of the chiming effect.

I'm reluctant to leave Kantju, feel I could linger here forever, but if I want to complete the circuit I should get moving. I head out on a thirty-minute walk along the north side of the Rock, mostly in the sun, and, to my surprise, encounter no other people. It's hot, but not such a scorcher as to drive people away. Clearly, few visitors make the complete Base Walk. Along the way I pause to watch a pair of little eagles (*Iṯaṯuṟa*) circling and diving with each other. The rock must create some interesting updrafts for the birds to play upon. Feathers from these birds play a role in male initiation ceremonies that take place on the eastern side of Uluṟu, yet another chapter in this storied desert.

I gradually circle to the opposite side of the rock and eventually arrive at Muṯitjulu waterhole, the setting for another important Tjukurpa story. This tale concerns the battle between Liru (a venomous king brown snake) and Kuniya (a woma python). As I step back to face Muṯitjulu, I can see a long line of discolored rock on the right side, the presence of Kuniya, and a large crack in the rock on the left, Liru. The brochure explains that "In the beginning, the female Kuniya becomes angry with the lack of respect for proper behaviour shown by Liru. She attacks him in a great rage, and the battle becomes a disastrous encounter." This story reminds me to be on the lookout for snakes—the king brown is especially venomous—and I'm probably not as cautious as I should be, spending too much time looking up at the rock or peering at birds and too little time keeping an alert eye on the trail ahead.

As I follow the path that leads to the Muṯitjulu waterhole at the base of the rock I enter the landscape of another snake, Waṉampi, the water snake, one of the powerful Rainbow Serpents in the Tjukurpa of central and northern Australia. This snake, the brochure explains, "has the power to bring or stop the supply of water if it has any reason to be angry." Because of this power, the "Aṉangu always show respect for Waṉampi." A sign at the waterhole, for example, requests "Please do not swim or disturb this place in any way." I can imagine on really hot days the temptation would be strong to take a quick dip.

From here, the trail passes another large rock shelter filled with pictographs. The brochure explains: "*This place was the major camping location for Aṉangu at Uluṟu. Here, close to the waterhole, our families have lived in shelter from the sun, wind and rain since ancient times. Many of us who live in Uluṟu Kata-Tjuṯa*

National Park today, lived right here when we were younger."[20] I feel an uncanny disorientation when I realize that these locations that I might think of as being very old, relics of an ancient time, were until very recently lived in as homes. And local people continue to utilize many of the sacred locations for ceremony. The Tjukurpa might reach back into Time Immemorial, but it is still very much alive today, as it will be tomorrow. This is in fact one of the ideas of the Tjukurpa Dreamtime that Westerners like me, with our deeply ingrained sense of linear time, have a hard time grasping. The Tjukurpa exists outside of time and is as alive today as it was 50,000 years ago. The Tjukurpa beings were, are, and will continue to be present here. I feel, though, that even as I speak of the past, present, and future, of 50,000 years of time, that I am resorting to the vocabulary of linear time, revealing to myself that I am incapable of grasping the everwhen of the Tjukurpa.

From the rock shelter, the trail swings around Uluṟu's western face, entering the territory of the Lungkaṯa, the blue tongue lizard. The Tjukurpa story associated with Lungkaṯa describes how one day he chanced upon an emu that had been wounded by hunters, but, in violation of ethical protocol, he greedily "kept the emu to himself and lied to the other hunters when they asked him where it was." When these hunters discovered the deception they chased after Lungkaṯa, and the boulders I pass beside the trail today are the pieces of cooked emu he dropped as he was pursued.

As I read these tales at the sites where they occur—the story of the Mala, of Liru and Kuniya, the accounts of Waṉampi, the tale of Lungkaṯa—I'm struck by the fact that the conflict in each of the stories seems to hinge on the issue of respect and the violation of an ethical code. The plot is set in motion when someone is disrespectful, and dire consequences ensue: a monster chases the people, a deadly fight breaks out, life-giving water is withheld, a thief is pursued to his death. And I wonder if in selecting these tales to retell in this brochure the Aṉangu aren't trying to teach the tourists to be respectful of this place, to honor their wishes about how to engage with this landscape, suggesting both an environmental and a social ethic for proper behavior.

In the years since handback, the Aṉangu have been increasingly involved in the management of the park, including in the role of park rangers, and the community has gained authority over the many tour guides and travel companies that visit Uluṟu, requiring them to undergo training that emphasizes Aboriginal culture. Furthermore, since handback two significant additions have been made to the park. One is the Cultural Centre, which displays information about the Tjukurpa as the Aṉangu wish to tell it. There's also a cafe, Ininti, where I head after my hike for some refreshments. Also located at the Cultural Centre is the Maruku Art Gallery, where a local woman is painting, her canvas spread out on the ground as she patiently applies multiple dots of paint. Her painting is a visual rendering of one of the Tjukurpa tales, a sort of mythic map, a stylized rendering of place. I peruse the displayed works, buy a few, and the staff at the gallery packages them for shipment to my home on the other side of the planet. I have to admire the efficiency of the system.

The other addition to the park is a new trail, the Talinguṟu Nyakunytjaku Walk. This actually consists of two short sections, the Minymaku Walk, where signage discusses "women's business" involving the collection and processing of bush foods, and the Watiku Walk, where signs detail "men's business" such as toolmaking and hunting. This system of trails lies roughly a mile from Uluṟu and includes several raised platforms that provide spectacular views of the rock, especially at sunrise.

The existence of the art center and of the Talinguṟu Nyakunytjaku Walk suggests some of the ways the handback of Uluṟu, as well as similar handbacks, has served as catalysts, rippling far out to revitalize Indigenous languages, culture, arts, and ceremonial traditions. Aboriginal activist Kaleesha Morris (Gumbaynggirr, Dunghutti and Kulkalgal) argues that "language revitalisation is fundamental to land back."[21] Such handbacks likewise enable displaced Aboriginal people to reconnect with their traditional Country, renewing ancestral connections and providing a locus to reconnect with family.

As noted earlier, one of the contentious issues at Uluṟu has been the climb. Since the 1985 handback the Aṉangu have put some limitations on the climb to the summit. For safety reasons, they would close it during risky weather such as high winds or extreme heat. Likewise, as a mark of respect, they would close the climb if an Aṉangu elder had recently passed away. And of course, the polite *request* not to climb has been in effect throughout this period. However in October 2018, several years after my last visit to Uluṟu, the ten-member board of directors reached a unanimous decision to ban all climbing of the rock the following October, giving people who really, really wanted to climb it a year's notice. And so in October 2019 the climb was officially closed. The handrail along portions of the trail has also been removed. As one might imagine there was considerable opposition to this decision. A news story, for example, notes that "critics of the climbing ban have pointed to the fact that traditional owners originally allowed the practice to continue after title to Uluṟu was handed back to the Aṉangu in 1985." However, Sammy Wilson, a senior custodian at Uluṟu, responded by explaining that at the time of handback, "there was a lot of anxiety. A lot of pressure was put on traditional owners to keep it open, and they were worried about that." Wilson concluded, however, that during the intervening years the Traditional Owners "didn't lose sight of the sacred law, and they wanted to teach people about Uluṟu's importance."[22] This decision to close the climb in spite of substantial opposition strikes me as a sign of increasing confidence, authority, and appreciation of the meaning of sovereignty among the Aṉangu.

When the park was handed back in 1985, some people rooted for the Aṉangu to fail. The conservative government of the Northern Territory strongly opposed the handback, as did segments of the tourism industry and some members of the general Australian population. This put the Aṉangu joint-management scheme under intense scrutiny. If it failed, if the park deteriorated, if tourism declined, if the ecology suffered, it would be a setback not just for the joint management of Uluṟu, but for the very idea of handback and joint management elsewhere.

Likewise, because of its prominence and historical significance, the Uluṟu handback and joint management arrangement has been intensively studied by ecologists, sociologists, political scientists, anthropologists, and a host of other sorts of academics. In an assessment of the first 15 years, research ecologist Graham Griffin concluded that the overall project had been quite successful, although there were still important areas for improvement. He noted that

> the Uluṟu model of ownership and joint management has succeeded in many areas of shared resource use, knowledge, and income generation. The park is managed to the highest world standards and recognized as the most important symbol of Australia. The traditional Aboriginal owners of the park have shared in some of the benefits, but in doing so have exposed their culture to high pressures and interests.

Griffin explained that in spite of their crucial role in managing the park, the Aṉangu "have profited little from the massive tourism industry that has besieged them and their homelands." He concludes that the "Aṉangu are engaged in a difficult task, protecting traditional concerns within the context of an inevitably dynamic culture, while under intense scrutiny as custodians of the national icon, and as exemplars of Aboriginal ownership, control, and joint management."[23]

More recently an advisory group commissioned by the federal government's Environment Minister assessed the joint management arrangements at three federally involved national parks, Uluṟu-Kata Tjuṯa, Kakadu, and Booderee. The report noted that many of the problems outlined by Griffin still remain. The primary concerns involve deteriorating relations between the joint management boards located at the parks and the central government authorities in Canberra. Summarizing the report, Cameron Gooley notes that after 30 years, many Aboriginal people are unhappy with the failure of the joint management arrangement to improve the socio-economic condition of their people to the degree they had anticipated. Due to what the Traditional Owners believe to be a failure to live up to promises, and what they consider to be a disrespectful attitude displayed by the federal government agency Parks Australia, trust has broken down, especially at Kakadu.[24] The report argues that "traditional land and biodiversity management practises have been given insufficient place in park management and Western methods have predominated." The advisory report cited 24 areas for improvement, including:

- Changing the culture in Parks Australia to recognize Indigenous ownership and cultural authority of the parks and to reverse the bureaucratic Canberra-focused management model;
- Moving more decision-making authority away from Canberra and to the parks;

- Improving employment and training opportunities for Traditional Owners on Country; and
- Reviewing funding arrangements to address deficits and make plans to move towards financial sustainability.

It is notable that all of these recommendations argue for more, not less, Aboriginal control of the parks. They also propose that the Aboriginal communities receive a greater share of the proceeds from tourist visits. While it is discouraging to see these lingering problems, it is heartening that even as they call for improvements, such periodic assessments generally support the positive benefits of the joint-management system.

In 2015, on the 30th anniversary of the handback, Traditional Owner Reggie Uluru, who was present at the original handback and who has worked as a ranger at the park, summed up the feelings of many:

> The land was being returned to its original owners, so we were happy. Long ago An̲angu were afraid because they were pushed out of their lands. And because of that, An̲angu left. But now a lot of people want to come back. That's good. It's our place, our land.[25]

That is, while much work remains to be done to improve the joint-management system, it has nevertheless proven to be an overall benefit to the An̲angu.

<◇>

Roughly 40 percent of Australia is currently under some form of Native title or other Indigenous land rights status, mainly in arid regions. While handbacks and landbacks can and do occur in many biomes, it is not by chance, I think, that so many have occurred in desert landscapes. The settler-colonial environmental imaginary that disparages deserts as wastelands can, in such cases, be exploited to advantage. Certainly, it's politically easier to transfer ownership of some "God-forsaken wasteland" than to transfer, say, a green, well watered, forest or a fertile, agriculturally rich prairie.

In the United States, many parks and other protected public lands are also located in arid regions. These include not just Park Service land but also Forest Service and Bureau of Land Management properties. Locations such as Grand Canyon National Park, Pipe Springs and Bears Ears National Monuments, and Chaco Culture National Historic Park, for example, would seem ideally situated for transfer to Native ownership and could well serve as test cases for the sort of federal landback process advocated by David Treuer. (Indeed as I write this the US Forest Service, the Bureau of Land Management, and the Bears Ears partner tribes—Hopi, Navajo, Ute, and Zuni—are negotiating a joint management agreement for the monument.)[26] Each location presents its own set of challenges, to be sure, but with mutual goodwill such difficulties should be solvable. Americans, both Native and settler, would be wise to study long existing handback efforts such as the one at Ulur̲u to gain ideas, strategies, and to identify potential problems.

Without waiting for federal government action, numerous other small-scale landback projects are currently taking place in the United States, mainly promoted by private citizens or nonprofit organizations. For example, in 2018 the Nature Conservancy of Nebraska donated 440 acres to the Ioway Tribe of Nebraska and Kansas for the development of the Ioway Tribal National Park. In December 2021, the Save the Redwoods League of Northern California returned 523 acres of ancient coastal redwood forest to the InterTribal Sinkyone Wilderness Council, a consortium of ten tribal nations in the region committed to "acquisition, lasting protection and cultural care for the lands and waters of our traditional territories."[27] In April 2022 the nonprofit Chesapeake Conservancy brokered a deal with the United States Fish and Wildlife Service and the Bureau of Indian Affairs to transfer 465 acres along the Rappahannock River in Virginia to the Rappahannock Tribe. Such efforts continue. The organization Landback, for example, is a project of the NDN Collective of Rapid City, South Dakota, a philanthropic organization dedicated to promoting Indigenous struggles. Landback's cornerstone effort is organizing for the return of all public lands in the Black Hills to the Lakota Nation.[28]

As I experienced during my visit to Ulur̲u, with the welcome sign, the brochures and trail signage, the Art and Culture center, the new culturally informed trail network, the An̲angu are telling their own stories their own way on their own land. There is no reason Native Americans should not be afforded the same right and opportunities.

Notes

1 In the accepted orthography of the An̲angu languages of central Australia, the retroflex—an underlined letter—is a standard feature, though not always employed in common usage. It is used with the letters "t," "l," "n," and "r." When used beneath an "n," as in "An̲angu," it signals a slight "r" sound, making the pronunciation "Arnangu." When used beneath an "r," as in "Ulur̲u," it indicates the standard English pronunciation of "r." When absent, the "r" would be pronounced liked the rolled "r" sound (similar to a "d") of many European languages, such as Spanish, Italian, or Polish.

2 David Treuer, "Return the National Parks to the Tribes," *The Atlantic*, May 2021, https://www.theatlantic.com/magazine/archive/2021/05/return-the-national-parks-to-the-tribes/618395/.

3 For an early discussion, see Mark David Spence, *Dispossessing the Wilderness: Indian Removal and the Making of the National Parks* (Oxford University Press, 1999).

4 See, for example, Joe Fox, Lauren Tierney, Seth Blanchard, and Gabriel Florit, "What Remains of Bears Ears," *Washington Post*, April 2, 2019, https://www.washingtonpost.com/graphics/2019/national/bears-ears/

5 The term "landback" was coined in 2018 by artist Arnell Tailfeathers (Blackfoot Confederacy) as part of a social media campaign to protest the Canadian government's complicity with pipeline construction. See, for example, Quinn Smith, Jr., "What Is Landback?" https://sites.duke.edu/thewellianmag/2021/11/08/what-is-land-back/.

It quickly caught on as a useful term to encompass a range of efforts at restoring Indigenous sovereignty over stolen lands.

6 Local and state politicians in Utah have long opposed the designation of federal land in their region as national monuments, parks, and wilderness areas. They have continued that opposition in regard to Bears Ears National Monument. For a succinct discussion, see Jon Kovash, "The Argument for and Against Bears Ears National Monument," *Utah Public Radio*, https://www.upr.org/news/2017-01-10/part-1-the-argument-for-and-against-bears-ears-national-monument.

7 Mount Olga was named by the explorer Ernest Giles in 1872, which he saw from a distance. It is a remarkable rock formation roughly 20 miles to the west of Uluṟu and is referred to as "Kata Tjuṯa," meaning "many heads" by the A<u>n</u>angu. Though appearing similar, it is composed of an entirely different rock formation from Uluṟu. Both are part of the current park complex.

8 Because of its low population and limited tax base, the Northern Territory has yet to attain statehood under the Australian Constitution. This is one reason the handback was able to proceed as an arrangement between the federal government and the A<u>n</u>angu in spite of the objections of the government of the Northern Territory.

9 In Australia "Traditional Owners" has become a common term to refer to the Indigenous people of an area. It is usually capitalized as a proper noun.

10 Details of this complex legal process can be found in "Uluru (Ayers Rock) National Park and Lake Amadeus: Luritja land claim: Report by the Aboriginal Land Commissioner, Mr. Justice Toohey to the Minister for Aboriginal Affairs and to the Minister for Home Affairs," Australia. Office of the Aboriginal Land Commissioner, Canberra: Australian Government Publishing Service, 1980. See also Graham Griffin, "Welcome to Aboriginal Land: Anangu Ownership and Management of Uluru-Kata Tjuta National Park," *Conservation and Mobile Indigenous Peoples: Displacement, Forced Settlement and Sustainable Development*, edited by Dawn Chatty and Marcus Colchester (Bergham, 2002), 362–376.

11 See, for example, https://parksaustralia.gov.au/uluru/files/nature/UKTNP-A4Factsheet-AnanguFire-Dec_2020_web-ready_FA.pdf

12 Dermot Smyth outlines the details of the joint management agreement in his article "Joint Management of National Parks in Australia" in Baker, Richard, Jocelyn Davies and Elspeth Young, editors. *Working on Country: Contemporary Indigenous Management of Australia's Lands and Coastal Regions* (Oxford University Press, 2001).
PARK JOINT MANAGEMENT

Aboriginal use of Park resources

In all jointly managed national parks discussed above, the rights of Aboriginal traditional owners to occupy and use the parks are recognised and protected in legislation, and/or leases agreements, and/or plans of management. On these Aboriginal lands, the rights of traditional owners to continue to hunt, fish, and gather are regarded as essential to maintenance of culture and identity. Aboriginal people also generally regard the use of traditional resources as part of the practice of caring for their traditional country. However, several mechanisms exist in each Park to balance this right to use resources with the obligation to protect biodiversity and other natural resources. These include:

- The application of Aboriginal law and tradition, which imposes restrictions on who can hunt (and fish and gather), what species can be hunted, where hunting can occur, and so on;
- The restriction of the right to hunt, fish, and gather to members of local Aboriginal people—there is no general right for all Aboriginal people to access, use, and occupy the park;
- Obligations imposed on the board and the conservation agency by legislation, lease agreements and plans of management to protect biodiversity and other natural values of the park;

- Similar obligations to protect the cultural values of the park; those values include traditional hunting, fishing, and gathering, which in turn must be sustainable in order to be protected;
- The power of boards to regulate Aboriginal hunting, fishing, and gathering if required.

The most recent Uluru-Kata Tjuta National Park joint-management plan can be found here: https://www.awe.gov.au/sites/default/files/documents/management-plan.pdf.

13 For a good discussion of Indigenous Protected Areas see Kim Mahood, "Country Needs People," *The Monthly*, July, 2017, https://www.themonthly.com.au/issue/2017/july/1498831200/kim-mahood/country-needs-people#mtr. For a more scientifically oriented discussion see Davies, J., R. Hill, F. J. Walsh, M. Sandford, D. Smyth, and M. C. Holmes. 2013, "Innovation in management plans for community conserved areas: experiences from Australian indigenous protected areas," *Ecology and Society* 18, no. 2: 14. http://dx.doi.org/10.5751/ES-05404-180214.

14 For example, in October 2021 the iconic Daintree Rainforest in coastal Queensland was returned to the Kuku Yalanji Traditional Owners and will be jointly managed as a park with the government of Queensland. Shortly afterward, in April of 2022, the Warlibirri National Park was established in conjunction with the Gooniyandi people on the Margaret River in the Kimberley region of Western Australia. https://www.abc.net.au/news/2022-04-28/new-national-park-kimberley-welcome-traditional-owners/101021718

15 For a detailed look at the motivations for those who choose to climb Uluru, see Hannah Hueneke's thesis "To Climb or Not to Climb: 'The Sacred Deed Done at Australia's Mighty Heart,'" Australia National University, Canberra, 2006. www.nintione.com.au/resource/HHthesis.pdf.

16 A brief definition of Tjukurpa is provided in the brochure "An Insight into Uluru":

The Tjukurpa is the traditional Law that explains existence and guides daily life. Like religions anywhere in the world, the Tjukurpa provides answers to important questions. How as the world and all life created, and by whom? How do people fit into the total picture of life? What are the laws that sustain nature and all living things?

The Tjukurpa is the foundation for Anangu. It provides the rules for behavior and for living together. It is the Law for caring for one another and for the land that supports our existence.

17 "Uluru Kata-Tjuta National Park, "An Insight into Uluru: The Mala Walk and the Mutitjulu Walk," Canberra: Parks Australia and Mutitjulu Community Incorporated, 1990.

18 Barry Hill, *The Rock: Travelling to Uluru* (Allen and Unwin, 1994), 154–155.

19 Robert Layton, *Uluru: An Aboriginal History of Ayers Rock* (Australian Institute of Aboriginal Studies, 1986), 5.

20 Italics in original.

21 Kaleesha Morris, "Land Back and Language," *Common Ground*, March 9, 2022. https://www.commonground.org.au/learn/land-back-and-language.

22 Peter McCutcheon, "Uluru Custodian Says It Is Time Climb Closure Critics 'Learn about the Way We See It,'" *ABC News*, July 22, 2019. https://www.abc.net.au/news/2019-07-22/uluru-climb-closure-chance-to-learn-rom-traditional-owners/11335086.

23 Graham Griffin, "Welcome to Aboriginal Land," *Conservation and Mobile Indigenous Peoples: Displacement, Forced Settlement and Sustainable Development*, edited by Dawn Chatty and Marcus Colchester (Berghan, 2008), 371. https://www.jstor.org/stable/j.ctt1btbx2j.27.

24 Cameron Gooley, "Deplorable State': Dysfunctional Relationship between Parks Australia and Traditional Owners." *Sydney Morning Herald*, March 9, 2022. https://www.smh.com.au/politics/federal/deplorable-state-dysfunctional-relationship-between-parks-australia-and-traditional-owners-20220309-p5a33t.html.

25 National Museum of Australia, "Handback of Uluṟu to the Aṉangu People," April 14, 2020, https://www.nma.gov.au/defining-moments/resources/uluru-handback-Anangu.
26 "BLM, Forest Service and Five Tribes of the Bears Ears Commission Commit to Historic Co-management of Bears Ears National Monument," June 21, 2022. U.S. Department of the Interior, https://www.doi.gov/pressreleases/blm-forest-service-and-five-tribes-bears-ears-commission-commit-historic-co-management.
27 Intertribal Sinkyone Wilderness Council, https://sinkyone.org/about.
28 For information on the Ioway park, see Paul Hammell, "Small Tribe to Create Sprawling 'Tribal National Park' Southeast of Rulo," *Omaha World-Herald*, November 29, 2020, https://omaha.com/eedition/sunrise/articles/small-tribe-to-create-sprawling-tribal-national-park-southeast-of-rulo/article_6235fc55-a67e-522f-97b4-d7d944fc6e3f.html. For details on the Rappahannock transfer, see Sarah Kuta, "Ancestral Homeland Returned to Rappahanock after more than 350 Years," *Smithsonian Magazine*, April 5, 2022, https://www.smithsonianmag.com/smart-news/ancestral-homeland-returned-to-rappahannock-tribe-after-more-than-350-years-180979863/. For information on Landback, see https://landback.org/. For examples of smaller-scale private efforts see *Reconciliation Rising*, https://www.reconciliationrising.org/ as well as Margaret D. Jacobs, *After One Hundred Winters: In Search of Reconciliation on America's Stolen Lands* (Princeton University Press, 2021).

Bibliography

Davies, J., R. Hill, F. J. Walsh, M. Sandford, D. Smyth, and M. C. Holmes. "Innovation in Management Plans for Community Conserved Areas: Experiences from Australian Indigenous Protected Areas." *Ecology and Society*, 18, no. 2 (2014): 14. http://dx.doi.org/10.5751/ES-05404-180214.

Fox, Joe, Lauren Tierney, Seth Blanchard, and Gabriel Florit. "What Remains of Bears Ears." *Washington Post*, April 2, 2019. https://www.washingtonpost.com/graphics/2019/national/bears-ears/.

Gooley, Cameron. "Deplorable State': Dysfunctional Relationship between Parks Australia and Traditional Owners." *Sydney Morning Herald*, March 9, 2022. https://www.smh.com.au/ politics/federal/deplorable-state-dysfunctional-relationship-between-parks-australia-and-traditional-owners-20220309-p5a33t.html.

Griffin, Graham. "Welcome to Aboriginal Land: Anangu Ownership and Management of Uluru-Kata Tjuta National Park." *Conservation and Mobile Indigenous Peoples: Displacement, Forced Settlement and Sustainable Development*, edited by Dawn Chatty and Marcus Colchester, 362–376. Bergham, 2002. https://www.jstor.org/stable/j.ctt1btbx2j.27

Hammell, Paul. "Small Tribe to Create Sprawling 'Tribal National Park' Southeast of Rulo." *Omaha World-Herald*, November 29, 2020. https://omaha.com/eedition/sunrise/articles/small-tribe-to-create-sprawling-tribal-national-park-southeast-of-rulo/article_6235fc55-a67e-522f-97b4-d7d944fc6e3f.html.

Hill, Barry. *The Rock: Travelling to Uluru*. Allen and Unwin, 1994.

Hueneke, Hannah. "To Climb or Not to Climb: 'The Sacred Deed Done at Australia's Mighty Heart.'" PhD diss., Australia National University, 2006. www.nintione.com.au/resource/HHthesis.pdf.

Jacobs, Margaret D. *After One Hundred Winters: In Search of Reconciliation on America's Stolen Lands*. Princeton University Press, 2021.

Kovash, Jon. "The Argument for and Against Bears Ears National Monument." Utah Public Radio. https://www.upr.org/news/2017-01-10/part-1-the-argument-for-and-against-bears-ears-national-monument.

Kuta, Sarah. "Ancestral Homeland Returned to Rappahanock after more than 350 Years." *Smithsonian Magazine*, April 5, 2022. https://www.smithsonianmag.com/smart-news/ancestral-homeland-returned-to-rappahannock-tribe-after-more-than-350-years-180979863/.

Layton, Robert. *Uluru: An Aboriginal History of Ayers Rock*. Australian Institute of Aboriginal Studies, 1986.

Mahood, Kim. "Country Needs People." *The Monthly*, July, 2017. https://www.themonthly.com.au/issue/2017/july/1498831200/kim-mahood/country-needs-people#mtr.

McDutcheon, Peter. "Uluṟu Custodian Says It Is Time Climb Closure Critics 'Learn about the Way We See It.'" *ABC News*, July 22, 2019. https://www.abc.net.au/news/2019-07-22/uluru-climb-closure-chance-to-learn-rom-traditional-owners/11335086.

Morris, Kaleesha. "Land Back and Language." *Common Ground*, March 9, 2022. https://www.commonground.org.au/learn/land-back-and-language.

National Museum of Australia. "Handback of Uluṟu to the Aṉangu People." April 14, 2020. https://www.nma.gov.au/defining-moments/resources/uluru-handback-anangu#:~:text=Handing%20back%20Uluru,deeds%20to%20Uluru%E2%80%93Kata%20Tjuta.

Parks Australia and Mutitjulu Community Incorporated. "Uluṟu Kata-Tjuṯa National Park, An Insight into Uluṟu: The Mala Walk and the Muṯitjulu Walk." Parks Australia, 1990.

Smith, Quinn, Jr. "What Is Landback?" *The Wellian Magazine*, November 8, 2021. https://sites.duke.edu/thewellianmag/2021/11/08/what-is-land-back/.

Smyth, Dermot. "National Parks in Australia." In *Working on Country: Contemporary Indigenous Management of Australia's Lands and Coastal Regions*, edited by Richard Baker, Jocelyn Davies, and Elspeth Young, 75–91. Oxford University Press, 2001.

Spence, Mark David. *Dispossessing the Wilderness: Indian Removal and the Making of the National Parks*. Oxford University Press, 1999.

Toohey, Justice. "Uluru (Ayers Rock) National Park and Lake Amadeus: Luritja Land Claim: Report by the Aboriginal Land Commissioner, Mr. Justice Toohey to the Minister for Aboriginal Affairs and to the Minister for Home Affairs." Australian Government Publishing Service, 1980.

Treuer, David. "Return the National Parks to the Tribes." *The Atlantic*, May 2021. https://www.theatlantic.com/magazine/archive/2021/05/return-the-national-parks-to-the-tribes/618395/.

United States Department of the Interior. "BLM, Forest Service and Five Tribes of the Bears Ears Commission Commit to Historic Co-Management of Bears Ears National Monument." June 21, 2022. https://www.doi.gov/pressreleases/blm-forest-service-and-five-tribes-bears-ears-commission-commit-historic-co-management.

12

BORDERLANDS ECOLOGIES

Informed Imaginaries in *Signs Preceding the End of the World* and *Encantado: Desert Monologues*

Celina Osuna

In North America, as in most other places on this planet, deserts are complicated by the construction of borders, the impacts of which have disproportionately affected Indigenous communities and migrants for centuries. For example, the US-Mexico border took most of its present-day shape with the signing of the Treaty of Guadalupe Hidalgo in 1848. As part of this agreement, Mexico ceded nearly half of its land to the US—land that was previously colonized by Spain and therefore stolen from Indigenous peoples. The rest of the line that takes its present-day shape separating the US and Mexico was created as part of another agreement between the two nations called the Gadsden Purchase (also known as the Treaty of Mesilla). This went into effect in 1854 and was fueled by a drive for land that could be used for the southern transcontinental railroad, which had its own consequences on people in the region.

A lesser acknowledged repercussion is how this border also imposed a divide of the O'odham people's land, a division that has affected their communities throughout the twentieth and twenty-first century especially as increasingly enforced "immigration laws prevent the O'odham from crossing it freely."[1,2] Because nearly all of the present-day US-Mexico border is situated in desert places, research about its evolution, and especially the harm done to people and land by policies invented for surveilling and policing it, engages an understanding of storied deserts that inherently goes beyond aesthetics and involves an exploration of human and more than human relationships.[3] Storied deserts is an approach through which one recognizes and embraces the value of desert abundance, a term I use to refer to the cultural and biological diversity of human and nonhuman desert communities and inhabitants. It is a concept that takes seriously the importance of stories and storied matter alike, especially as deserts are often misrepresented as devoid of both.

This chapter, taking up what Wald et al. have called "Latinx environmentalisms," engages two contemporary literary texts that showcase the abundance, or

DOI: 10.4324/9781003397212-16

multiplicity and possibility, of desert places.[4] As what I call informed imaginaries, Yuri Herrera's novella *Signs Preceding the End of the World* (2015) and Pat Mora's poetry collection *Encantado: Desert Monologues* (2018) are stories rooted in the realities of the so-called Desert Southwest, though they make use of the speculative to show the nonexhaustive range of the literary desert. Informed imaginaries provide a way to think with and beyond what has been and what is. They are also crucial for imagining otherwise and even creating conditions for what will be. Herrera's *Signs* is a brief but profound reimagining of a migrant narrative in the desert borderlands, a setting that simultaneously serves as the nine levels of the Aztec underworld, Mictlán. Mora's *Encantado*, much lighter in tone than Herrera's book, explores the intimacies and interrelationships of people and place in the eponymous fictional town. Through their careful attention to language and environmental or nonhuman agency, both texts invite readers to think differently about desert places as abundant, full of history, violence, wonder, and hope—as storied deserts.

A Desert Odyssey

Originally published in Spanish in 2009, Yuri Herrera's *Signs Preceding the End of the World* was translated to English by Lisa Dillman in 2015.[5] The novella's central protagonist is a young woman named Makina who sets out on a quest to find her brother three years after he has migrated to the North, or "the other side."[6] Makina is from a place simply named the Village, where she is a telephone switchboard operator. The community values her work because she is trilingual, fluent in what the book calls "latin," "native," and "anglo," serving as a bridge between people—a role she, like many Latinx people, finds herself in often, no matter how far away from her hometown she may be. And by the end of Herrera's story, she ends up worlds away from where she started. Indeed, *Signs* follows Makina as the heroine of a desert Odyssey as she navigates the book's nine chapters, each paralleling the nine levels of Mictlán, as they describe experiences that reflect the borderlands ecologies of the US and Mexico as Makina encounters and embodies them.[7] Her journey includes migrant experiences of crossing the border via a river, being apprehended by law enforcement, and the formation of transborder identity. Structurally and thematically, these layers of meaning make *Signs* a palimpsest that contains familiar and unfamiliar ways of thinking about the deserts of the US-Mexico borderlands.

This section illuminates how *Signs* experiments with ways of storytelling, new and old, to produce an original lens for understanding contemporary Latinx and borderlands literature. Literary scholar Marcelo Rioseco writes that the text

> constructs a new space in which the border is not between two distinct geographical areas, between two countries, but rather between reality and myth.... Rewriting that is not a copy but a model, a guide; its originality comes from its very antiquity.[8]

This 'rewriting' reveals to readers the possibilities of abundance in the literary desert as well as the realities of the US-Mexico borderlands; desert abundance is a key component of storied deserts, as they both work to construct a fuller conceptualization of desert places as complex and diverse. It is especially Herrera's exploration of language that provides fresh ways of perceiving desert places, even as it mirrors the mythic. Rioseco attests, "without a doubt, what distinguishes and isolates Herrera from other border writers—or from narco literature in general—are not the subjects he addresses, but rather the language with which he addresses these subjects."[9] Indeed, the author's language defamiliarizes popular views of deserts and their stories, connecting the familiar to the unfamiliar—the new, the possible, the unknown or the not-yet-known. As an informed imaginary, *Signs* tells a robust and epic story in a way that negates the voiding of deserts by way of world-building.

One of the most obvious ways that Herrera's language shapes the reader's relationships to its environments is through naming or even lack of naming. There is enough context in *Signs* to infer that "the North" could represent the United States and "the South," including "the Big Chilango," "the Little Town," and "the Village," could represent Mexico and places within it. One can also take an educated guess that when the narrator mentions Makina speaking "latin," "anglo," and "native," that those correlate to Spanish, English, and an unspecified Indigenous language, respectively. Translator Dillman elaborates that she was mindful in her translating practice to use "an English that was geographically non-explicit," while at the same time leaving in words that would suffer from being translated, leaving those in to remind readers that this is a translated text.[10] Thus, cultural traces contribute to the story's mythical dimension, as Makina's migration to the North can simultaneously be interpreted as her journey through the underworld of Mictlán. Without the overt use of "English" or "Spanish," without naming "the United States" or "Mexico," the fictional world built by *Signs* is informed by but not confined to ready associations with nonfictional geopolitical relationships.

As an element of identity and sense of place, language in *Signs* is always navigated alongside the land. Herrera drives this point home when the narrator, in parentheses, describes the "latin-anglo dictionary" that Makina has packed:

> (those things were by old men and for old men, outdated the second they left the press, true, but they still helped, like people who don't really know where a street is and yet point you in the right direction).[11]

Makina's trilingualism enables her to think and act between different languages and interlocutors. This is a familiar way of being for her; she is successful as a switchboard operator in the Village because of her ability to speak three languages and "keep quiet in all three, too."[12] So, naturally, when she reaches "The Place

Where the Wind Cuts Like A Knife" and observes people in the North speaking "an intermediary tongue," she

> instantly warms to [it] because it's like her: malleable, erasable, permeable; a hinge pivoting between two like but distant souls.... More than the midpoint between homegrown and anglo their tongue is a nebulous territory between what is dying out and what is not yet born.[13]

The relationship between "homegrown" and "anglo" tongues reflects a familiar reality that many people in the US-Mexico Borderlands embody and becomes a realm of possibility through a co-constitutive way of sharing stories that is predicated not on either/or but by both/and:

> Using in one tongue the word for a thing in the other makes the attributes of both resound; if you say Give me fire when they say Give me light, what is not to be learned about fire, light, and the act of giving? It's not another way of saying things; these are new things. The world happening anew, Makina realizes: promising other things, signifying other things, producing different objects.[14]

Instead of insisting on the limits of language, Herrera has much more interest in exploring its bounties, its reverberations. Herrera's desert world is one where the potential for abundance and collaboration outweighs the existence of dearth or desolation, even as they all coexist. The people Makina meets have come to the North and navigated it with a combination of both "ancient memory and the wonderment of new people."[15] Throughout *Signs* there is an emphasis on the new or the not-yet, especially as it relates to what has been before.

Versing

As a speculative tale of transformation and translation, *Signs Preceding the End of the World* is Makina's desert epic—one that requires new language to think differently about migration's potential as a mythic mode of movement. In her translator's note, Dillman explains the range of texts, spanning from Lewis Carroll's *Alice in Wonderland* and McCarthy's *The Road* to books about Aztec mythology, that she used as references to build a sense of style and nonstandard language to suit Herrera's novella. Among the tasks at hand was rendering *jarchar*, "the novel's most talked about neologism" into English.[16] She writes about the word's meaning and etymology:

> Within *Signs*, it means, essentially, "to leave." The word is derived from *jarchas* (from the Arabic *kharja*, meaning exit), which were short Mozarabic verses or couplets tacked on to the end of longer Arabic or Hebrew poems written in

> Al-Andalus, the region we now call Spain. Written in the vernacular, these lyric compositions served as a sort of bridge between cultures and languages. And on one level *Signs* is just that: a book about bridging cultures and languages.[17]

The poetic roots of *jarchar* were key to Dillman's decision to use "verse" as the English translation of the word. Unlike with *jarchar*, which is new to Spanish language readers, English language readers will have some familiarity with the signifier "verse." Dillman explains,

> used in context [to verse] is easily understood, and has the added benefits of also being a noun-turned-verb, a term clearly referring to poetry, and part of several verbs involving motion and communication (traverse, reverse, converse) as well as the "end" of the uni-verse.[18]

To verse provides a new understanding of movement "in a way that explores and expands mental and emotional realms through the tool—or weapon, if the violence of Herrera's novel is any hint—of language," imbuing the story with an ever-present poetics that complements Makina's epic journey.[19]

Indeed, versing illustrates how a man named Chucho helps Makina cross the river and is "warm even though he'd only just versed from the water" and how, when Makina is shot as part of an altercation, "the bullet had entered and versed between two ribs."[20] Any time something or someone verses, the action is layered with language that expands the book to the scale of the mythic. Chucho isn't just some man coming out of the water to take Makina across the river, he "versed from the water," like a gatekeeper between two worlds, the North and the South. Makina not only survives the bullet wound, but she hardly feels any pain at all—a true heroine. To this end, when Makina receives the address where her brother might be, she takes on superhuman characteristics to expedite her journey: "Makina flew; she literally felt her feet not touching the ground, as if she could float, scissoring her legs until she found her brother and brought him home without setting foot on foreign soil again."[21] Her movement, and the way she navigates or verses from a place, is swift and lyrical, alluding to her heroic nature and to the messenger god Hermes, who has wings on his feet and crosses borders. While things and people alike can verse in this text, translator Dillman notes that Makina "is the character who most often 'verses,' as well as the woman who serves as a bridge between cultures, languages, and worlds."[22] In this way her plight is singular, even as it produces a connective tissue for everything else in the book. In other words, though Makina is not herself always bound to the same physical plight as others, the people and situations she encounters along the way highlight the histories and interrelationships of migrants crossing through or living in the desert borderlands.

Once the reader becomes familiar with how characters and objects verse from one point to another, the word and its poetics offer a speculative approach for engaging migration narratives of the US-Mexico borderlands and their storied deserts.

Rather than repeat well-worn and harmful rhetoric rooted in fear—terms like "illegal immigrants" or "aliens" that "swarm" or "flood" borders for example—versing expands such narrow views by inviting readers into a realm of wonder. In *Border Ecology: Art and Environmental Crisis at the Margins*, Ila Nicole Sheren writes that "the border becomes a truly generative site, one that, in its very thickness, creates a space for alternate modes of thinking, for a dismantling of power structures. Suddenly, it is the center that is hopelessly out of touch."[23] Makina's story, of crossing an inferred international border and simultaneously the borders between the levels of Mictlán, emphasizes possibility, not limitation; it's an invitation to think differently about movement and migration. Makina only travels to the North because her mother asked her to, and she takes very few things with her (a picture, a bar of soap, and some peanut brittle) an indication that in her mind, "she was coming right back."[24] Instead of a plot driven by someone desperate to survive by migrating north through a constant battle for resources like shade, water, and food, *Signs* offers an exploration of an alternative story, as Makina is largely uninhibited by the obstacles she faces. As such, the text overturns the logics of scarcity that usually dominate desert stories.

However, just because Herrera's novella is primarily concerned with Makina's journey doesn't mean the text erases or ignores the conditions of hardship and violence imposed upon other migrants who come from the South. On the contrary, it interrogates them through Makina, exposing the biopolitics, necropolitics, and geopolitics that are inextricably woven into the fabric of the borderlands. This includes aspects regarding patrol and surveillance that are clearly informed by the realities of the region, such as detention, deportation, and smuggling through the use of coyotes.[25] As Sheren puts it, "in the logic of any given center ... borders require maintenance, patrol, walls, fences: a militarized guard."[26] Importantly, Makina's ability both to verse and to write verse enables her to thrive by confronting such "maintenance," and assisting other migrants whenever she can. In a scene where she and others are lined up on their knees next to one another and taunted by a cop, it is literally her poetry that saves them. Of the people there, all of whom "were or looked homegrown," the policeman spots that one of the men has a notebook, and he grabs it and announces to the group: "Poetry. Lookie here at the educated worker, comes with no money, no papers, but hey, poems."[27] He then tells the homegrown man to "write why you think you're up the creek, why you think your ass is in the hands of this patriotic officer."[28] The man is too nervous to write anything. Makina, sensing this, grabs the pencil and paper and begins to write instead. The cop begins reading it aloud in one of the most poignant passages of the book:

> We are to blame for this destruction, we who don't speak your tongue and don't know how to keep quiet either. We who didn't come by boat, who dirty up your doorsteps with our dust, who break your barbed wire. We who came to take your jobs, who dream of wiping your shit, who long to work all hours. We who fill your shiny clean streets with the smell of food, who brought you violence you'd

> never known, who deliver your dope, who deserve to be chained by neck and feet. We who are happy to die for you, what else could we do? We, the ones who are waiting for who knows what. We, the dark, the short, the greasy, the shifty, the fat, the anemic. We the barbarians.[29]

Makina's words—her verses—are so effective that the policeman finishes reading and just walks away from the whole scene, leaving Makina and the men to go free. She disarms his bigoted disposition with vivid imagery that in no uncertain terms demonstrates that she understands exactly what people like that "patriotic" policeman in the North think of her and other "homegrown" people from the South. Rather than have the cop relay this language as his own, Herrera has Makina write it out for him, defamiliarizing the power dynamic between an authority of the law and a migrant, and ultimately collapsing racist rhetoric that is unfortunately commonplace in public discussions of the borderlands.

Another striking example of defamiliarization used by the author to instigate the political power structures of the North is an allegorical description of baseball. While in the North, Makina finds an old man who guides her to a baseball stadium to hand over Mr. Aitch's package to Mr. P, a task she was given as part of her passage north. When she inquires about what the stadium is used for, he replies:

> They play, said the old man. Every week the anglos play a game to celebrate who they are.... One of them whacks it, then sets off like it was a trip around the world, to every one of the bases out there, you know the anglos have bases all over the world, right? Well the one who whacks it runs from one to the next while the others keep taking swings to distract their enemies, and if he doesn't get caught he makes it home and his people welcome him with open arms and cheering.[30]

Here baseball becomes much more than America's favorite pastime; deconstructed like this, the game becomes a revelation of US identity, war, and imperialism. The old man's explanation enlarges the playing field from a local to a global scale, and the bases don't need to be called anything else to suggest US military presence around the world.

Thus *Signs* declares that America's actual pastime is war, tapping into the reality of US militarization of desert places. Indeed, this is a long-standing practice that took especially deep roots in the wake of World War II. Recently, the film *Oppenheimer* (2023) has repopularized the Manhattan Project's development and subsequent testing of nuclear bombs in Los Alamos, New Mexico, in the 1940s, including the infamous first testing at the Trinity Site.[31] Notably Christopher Nolan's film does not address the displacement of Indigenous communities that occurred as a result of the military operation, though there are a couple of times the place is mentioned as Native land. It also fails to acknowledge the exploitative labor practices created by a regional uranium boom, where "from 1944 to 1986, nearly

30 million tons of uranium ore were extracted," subjected the miners and their families, the majority of whom were Diné (Navajo), to serious health conditions from radiation exposure, like lung cancer.[32] It was shortly after this that the area, called the Nevada Test Site (NTS) until 2010, began to carry out atmospheric and underground nuclear tests—over 900 of them from 1951 to 1992.[33,34] Literary scholar John Beck, who has written extensively on militarization of the American West, lists these projects and their "subsequent poisoning by fallout of people and land" along with "the internment of people of Japanese origin in the West in the immediate aftermath of Pearl Harbor; … the progressive military withdrawal from public use of millions of acres of Western land; … the plans for permanent storage of contaminated waste in the Western desert; the "war on drugs" and intensive policing of the international border with Mexico" as clear indicators for "ways in which the West has been at war."[35] This list makes clear that land so often branded as barren or empty is perceived by the national government as abundant with purpose and possibility. My argument in this chapter and in much of my scholarship is for individuals—who do not already—to realize this desert abundance, too. In doing so, there is more likelihood for environmental and social justice for desert communities around the world and the delicate ecosystems upon which they depend.

Elemental Immediacy

Beyond use value, storied deserts also take interest in arid places as agential. Well known for their extreme conditions, deserts are abundant with the nonhuman and more than human. Elements like light, water, wind, heat or even cold have, over millions of years, produced diverse geologies and biologies whose temporalities are difficult to truly apprehend. Herrera often oscillates between human and nonhuman scales in *Signs*, using the vehicle of fiction to reveal the land's agency. The book begins:

> I'm dead, Makina said to herself when everything lurched: a man with a cane was crossing the street, a dull groan suddenly surged through the asphalt, the man stood still as if waiting for someone to repeat the question and then the earth opened up beneath his feet.[36]

The sinkhole in this passage enacts what I call elemental immediacy—a sudden eruption of more than human intensities or forces. Such irruptions provide a glimpse at the overwhelming power of the elemental, confronting the limits of human perception. Events of elemental immediacy, tragically becoming more commonplace due to climate change, can pose serious threats to life, to infrastructures, and can trouble notions of predictability in terms of things like weather patterns, agricultural yields, and species population growth or decline. In the opening paragraph of *Signs*, the earth opens up in an act of elemental immediacy and "swallowed the man, and with him a car and a dog, all the oxygen around and even the screams of passers-by."[37] The list here is a taxonomy of things; the earth devours a person, an animal, an object, atmosphere, and sounds, indicating from the very beginning of the story that this is no passive or scenic landscape.

As a story about Makina's journey through the US-Mexico borderlands *and* the borderlands of the underworld, she quite possibly does die with the emergence of the sinkhole on the first page, a victim of its elemental immediacy. Taking into account the novella's basis on Mictlán, readers can interpret this event as the very first after her death, as "[i]t is likely that before undertaking this journey [to the underworld], the individual had to be devoured by Tlaltecuhtli, Lord/Lady of the Earth."[38] This reading of the book also helps explain how Makina verses through the violence of the text without it ever truly impeding her journey.

In the scene where Makina reaches the river, her guide Chucho takes her across on an inner tube. Like the rest of the novella, this passage can be interpreted as an example of borderland migration (across a river like the Rio Grande) *and* as movement through the levels of Mictlán. Evidence for the latter interpretation includes that "chucho" is a word in Spanish for dog (or mutt). It is a dog, specifically a xoloitzcuintle, that guides souls across the Apanohuaia river in the first level of the underworld. In either reading, during their crossing "an icy current began to push their feet away like a living thing, relentless," and an unforgiving quickness fills the moment.[39] The current overturns the pair, who are briefly stranded in the water before they reach the other side, and the tube is "swirling away … as if it had urgent business to attend to."[40] Each of these depictions, emphasized by swift motion, treats the river and the innertube like they have their own agendas that momentarily though abruptly entangle with Makina before they carry on.

In the chapter entitled "The Place Where the Hills Meet," the vastness of geological time suddenly becomes discernible to Makina. Though "first there was nothing," eventually she sees

> two mountains colliding in the back of beyond; like they'd come from who knows where and were headed to anyone's guess but had come together at that intense point in the nothingness and insisted on crashing noisily against each other, though the oblivious might think they simply stood there in silence.[41]

Makina sees noise where others see silence. Through her perception, which feels much like it's through the looking glass, the mountains crash into one another—something this informed imaginary allows us to witness that we wouldn't otherwise. Outside of fiction, this process of earth-shattering shifts takes tens of millions of years to occur; through fiction, it is arrested for our contemplation within the length of a paragraph. The land is emergent, active, and alive. Its tectonic movement, normally too imperceptible for one to notice, is boisterous. Ultimately, the environments of *Signs*, whether read as the US-Mexico borderlands or Mictlán, are eruptive and agential in ways that may be read as violent. To this end, one can argue they reflect the histories of the region, especially the colonial legacies of violence that continue to enact harm to this day. This is an important way to understand borderland ecologies, especially as the plural indicates the existence of other, less violent relationships that also contribute to its storied deserts.

A Desert Community

While Herrera's *Signs Preceding the End of the World* textures desert vastness with narrative language for the violence and possibility of migration, Pat Mora's *Encantado: Desert Monologues* creates a desert place with an atmosphere full of wonder and intimacy. Listing Thornton Wilder's *Our Town* (1938) and Edgar Lee Masters's *Spoon River Anthology* (1915) as its major influences, Mora's poetry collection is set in a fictional Desert Southwest town along a river. Like Herrera, Mora draws on literary traditions of the past in order to build a world informed by contemporary issues of identity and belonging in the borderlands. Mora, too, uses careful language to reveal desert abundance. The storied desert of Encantado, the town that gives the work its title, showcases the entanglements and encounters of its diverse community's human and nonhuman inhabitants. Its dramatis personae include men and women, people young and old, people from the town and people from out of town, people with different occupations, and people who are spouses, parents, or children. Over the course of the collection, readers become acquainted with the people living there, their relationships to each other and with the land itself. In reading *Encantado* one becomes familiar through intimacy with the many close relationships within its pages. Unlike in *Signs*, which keeps a pace so quick as to perform the constant motion of Makina's migration, intimacy is a key aspect of *Encantado* that contrasts narratives of trauma and violence with acts of love, healing, grief, and resilience. Working constructively, creatively against desert-as-void, Mora's poetry reimagines deserts through human intimacy with each other, the land, and even those whose spirits remain active community members.

The informed imaginary of *Encantado* exemplifies the realities and possibilities of living and belonging in a Desert Southwest town—present and future—often by providing stories that allude to the region's multicultural and complex past. Therefore, the collection frequently revolves around language, identity, and belonging. *Encantado* begins with "Señor Ortega," a title that broadly gestures towards the identity of its speaker (an elderly man of Latinx heritage) before the opening lines declare it outright: "I live in languages, Spanish, English— / and shoes, yes, old *zapatos*, their leather tongues."[42] Just two lines in, the cobbler hints at his occupation. He explains a bit about his shop, "Seventy years of rhythm, my hammer's *tap-tap-tap*, / the sewing machine's *whirr*," mostly business as usual, "except for that man / years ago who suggested I change my name / to improve my business. Crazy, some people."[43] Señor Ortega's identity as a cobbler and a Latino are linked by his name; something he would not give up despite what the man in this memory tells him. Though the encounter was a minor disruption to the decades-long rhythm of his work, it has clearly stayed with him for many years, and in the poem's final stanza the speaker reflects on his decision to ignore the racist request: "Gnarled as tree roots, these hands mend / flamenco taps. My name still on my sign."[44] Señor Ortega's story is one of resilience. His body has transformed over time, just as the worn-out shoes that he has spent his career repairing. But his name prevails on his sign, a lasting piece of identity that he refused to be ashamed of. The poem ends with a gesture to the future, "I pick up the next shoe or boot—like us, it needs patient attention and repair."[45] His final lines implicate an "us" that, like worn shoes, "needs patient attention and repair." The señor, even in his dotage, welcomes the challenge of healing and reconstructing, which takes time and requires hope. It also requires a strong community.

Naming

One of the ways Mora builds a sense of community in *Encantado* is through the use of naming. Most of the poems are named for their speakers, thus allowing them to tell their own stories from their own perspectives. Some poems even mention one another by name, and, over the course of the collection, the reader becomes familiar with the constellation of the town. Whereas Herrera's *Signs* uses names (and even non-naming) to conjure possibility through allegory, Mora's *Encantado* uses them to create a sense of intimacy with speakers and their stories. In my other work, *Desert Distortion*, I argue that "names are words with which we world"—they produce containers as well as possibilities—and Mora's imaginary town is no exception.[46] As "encantado/a" is a word used in Spanish when first meeting someone, the collection is in effect introducing readers to these speakers, providing intimate insight to the people and things that carry the burdens and blessings of the Southwest's complicated and multicultural history and amplifying the power and polyvocality of the region that is often ignored. Every monologue is a brief glimpse at someone's experience, shedding light on who they are, what they feel their purpose is, and their own connection to place. In her book *Porous Borders*, historian Julian Lim attests:

> though traditionally defined in terms of Mexican—Anglo-American relations, the borderlands were in fact traversed by a multiethnic and multiracial population that came together dynamically through work and play, in the streets and in homes, through war and marriage, and in the very act of crossing the border.[47]

Encantado's community, even as it centers Latinx identity, demonstrates that there is no homogeneous Desert Southwest; it presents a multiracial, multiethnic compilation of voices; there are over 40 characters whose experiences are captured into brief verse, and each poem is a piece of the mosaic of Encantado. Beginning *Encantado* with a story from a multilingual cobbler who, despite bigotry, refused to change the name—his name—on the shop's sign, emphasizes the incredible power that names have.

The kaleidoscopic nature of *Encantado* demonstrates desert abundance, as its pages are filled with so many different voices.[48] Rather than exhaustively relay each identity present within its pages (difficult, too, because the speakers have intersectional positionalities), the following is a list demonstrating the diversity of the town's community: "Eduardo" was once "A desert boy who always loved the West— / slow tales of cowboys round a fire, the stars" who gets a government job and imagines settlers in wagons crossing the desert from the seat of his car; "Phil" was a kid who was bullied, so he learned how to be a bully and carried this into his adult life where his amorality breeds wealth and he lives "concealed in expensive suits, cars"; "Lupe" recounts her joyous childhood as she stands working behind a counter, "erased by a uniform"; "Tuan" is an artist from Saigon who had to change from painting canvas to painting women's nails in order to provide for his family; "Cecilia" poetically describes "the pleasure of resting my hands on the hidden growing that began to precede me" and her journey of pregnancy and motherhood; "Rogelio" is a landscaper who conveys the moment he and his wife found their first home that wasn't a trailer, "felt its promise, / *para la familia*"; "Lucia" is a "body of Cuban memories" and a writer who relays the time her Papi grabbed a bull by the horns so that her and her mother could run away safely; "Don Arturo" sits at the river recounting his departed wife before heading home filled with pride that so many people of the town have chosen him to be their English teacher, "They come—grandmothers, uncles, young mothers—carrying notebooks, pencils, babies, hope; their bodies heavy with debts."[49] The speakers, and their stories, represent moments of celebration, struggle, mundanity, triumph, and grief. Reading their lives through these moments builds a sense of understanding for the characters—their names, families, histories, successes, losses, childhoods, livelihoods, and (dis) placements. Such intimate details also make evident the diversity of desert communities, obfuscating a homogenized or abstract conception of desert places which has contributed to practices of erasure and narratives of emptiness or lack.

Desert Enchantment

In addition to my own research, recent scholarship—such as Cynthia Radding's *Bountiful Deserts* (2022), Gary Paul Nabhan's edited collection *The Nature of Desert Nature* (2020), and Jada Ach and Gary Reger's *Reading Aridity in Western American Literature* (2021)—speaks increasingly to an important shift in the ways deserts in North America are perceived and represented as full.[50] For Radding, who discusses the lifeways of Indigenous desert peoples like the Tohono O'odham, the Comca'ac, and the Yoremem and Yoemem, "the peopling of the monte brings forth its abundance."[51,52] Indeed the case for many of the world's Indigenous desert communities is that their homelands, with which they have achieved sustainable relationships and from which their cosmologies emerge, have been colonized for the purposes of extraction, waste disposal and other pollution, or militarization while settler-colonial narratives characterize the very same arid places as barren and inhospitable to life. Nabhan speaks to a similar irony present in how deserts are defined "by most cultures in terms of what they lack rather than by the uniqueness of what they have."[53] The US-Mexico border culture that informs Pat Mora's collection is not one that views the desert through a lens of lack. In her earlier work, a collection of essays titled *Nepantla: Essays From the Land in the Middle* (1993), Mora asks

> How do we create space for ourselves to be ourselves, our multiple Latina, Hispana, Mexicana, Chicana selves? Space in ourselves to hear ourselves, space free of the pollution of bigotry and bias, space free of contemporary colonizers.... Space for play and dance.... The desert teaches us the value of space.[54]

I suggest that *Encantado* is an opportunity to learn from deserts and contemplate them as intimate settings storied with possibility.

Mora's imaginary desert town lives up to its name, a greeting and descriptor which translates to English as enchanted. As a collection, the poems convey a sense of enchantment, defined by Jane Bennett as a "mood of fullness, plenitude, or liveliness."[55] *Encantado*'s pages tell the stories of its human, nonhuman, and more than human inhabitants, transforming the desert into a place of wonder. This is especially evident in the poems "Encantado" and "November 2: El Día De Los Muertos," which express the role of spirits, or *ancestros*, for the town and those who live there. Both pieces portray a liveliness even in death that fills the town's atmosphere with love and abundance. In the first, "Encantado," the spacing and structure suggest a nonlinear, irregular but present connection between the spirits and the physical world as they interact:

> The last nights of the year,
> kind, departed spirits return
> to Encantado as stars,
> meander
> down dark streets and hallways,
> peer into windows,
> congregate around cribs,
> again leave glowing glints
> of themselves;
> intertwine with our dreams,
> shine on bare boughs,
> pines, and cactus spines.[56]

Encantado is comprised of its spirits as much as its streets and hallways, cribs, dreams, pines, and cactus spines—an enchanted desert assemblage where those who have lived and died before continue to visit and maintain relationships with the land and all who live on it. The collective spirits of the town are the speakers for "November 2…," a poem named for the day that Mexican culture celebrates the lives of those who have died. The piece begins with joy for their return, "such pleasure— / occasions without the fuss / of details."[57] They linger over the town's river, "gaze down at our humorous / rippling reflections" before visiting "familiar dusty streets and rooms / our town / through the sounds of the living / patting heads we love / comforting the attentive."[58] Just as in the previous poem, these spirits enliven the town, where they enact pleasure, humor, and compassion. Their admission to "comforting the attentive" suggests that those who stay connected to their deceased loved ones through attention like prayer or ritual will benefit from their visits each year, from the enchantments of this place.

For Bennett, a significant component of enchantment is sonority, which "refers to the aural effectivity that living, moving, snorting, weeping, laughing bodies possess. Sonority does not represent, for it does not operate via images or in the visual mode."[59] Throughout *Encantado*, Mora uses this "sonorous dimension of language," creating individual, intimate, and embodied experiences of place.[60] The poetry's sonority works against barrenness and instead constitutes a rhythm of the town. When the speakers, the people of Encantado, recount their stories with vivid sound imagery, it thickens the atmosphere of *Encantado*, the place and text, with sound and even silence. A return to Señor Ortega's shoe repair shop: "Seventy years of rhythm, my hammer's *tap-tap-tap*, the sewing machine's *whirr*, / quiet *thud*, the screen door—."[61] His septuagenarian rhythm is eventually interrupted, as we know, but in these lines the ensemble cast of his daily work life is easily pictured, emphasized by italicized onomatopoeias. Things are a lot quieter at home since Señor Ortega's wife passed away: "all I hear at home—TV, refrigerator."[62] The uneventful hum or buzz of appliances in contrast to the lively *whirr* and *tap-tap-tap* of the shop suggests an interesting contrast in rhythms. It's the array of rhythms and experiences that makes *Encantado* so interesting, that stories the town of Encantado. "Anna," remembers how her "Mother, a widow, a frightened, nervous / woman in her second country, / sits at her kitchen table, begins / her practiced meditation— / her hands rhythmically clean beans, / the familiar easing into her body."[63] Just like Señor Ortega, Anna's mother has an embodied individual rhythm when she uses her hands to practice something familiar.

The silences of *Encantado* also contribute to its rhythms and are just as important as its sounds; they attest to the power of absence rather than dismiss it as a lack. This point is especially relevant where deserts are concerned because, as this project on storied deserts discusses frequently, the empty desert is still a very prominent, though problematic depiction. In "Rita," the speaker reminisces about being raised by her grandmother, who she calls Mama. She remembers winter mornings when Mama "placed me in a pool of sunlight / on her well-swept dirt floor. / The wood stove and desert sun warmed / our silence."[64] The silence between Rita and Mama, between the stove and the sun, is warm, full of things Mama doesn't know how to say. Here, the desert atmosphere diffuses tension and exchanges it for what

Rita reflects is a wordless comfort, "maybe teaching me / to live alone"; they need not speak, only feel.[65] In "Stella," the speaker is "fretting again about words" and asks readers, "Can the doctor, my patient's son, hear / the pauses—implications?"[66] Stella hopes that her silence will speak for her and tell the man his mother is dying. Her hesitation supplants the need for words while conveying the message anyway. Mora makes tangible the silences, highlighting their ability to carry meaning and contesting the very equation of silence with emptiness.

Through sounds and silence, the people of Encantado are part of a community whose distinct rhythms contribute to a larger rhythm that permeates the text in the form of the river, el río, who serves as the town's most ancient storyteller. We already know that spirits come and linger to see themselves in its waters each year, but the river is important for the living as well. In "Amy," the speaker says: "We listen to the persistent river, / the town's balm."[67] Confirming its healing characteristics, many of the people of Encantado address the river directly. Don Arturo, in reference to his departed wife, asks, "Remember, old rio, how we would picnic here, / holding hands year after year?" and Cecilia, in reference to the baby in her womb, says, "My tenant and I strolled by you— / water within—and you, / rippling by carrying sunlight."[68] The speaker of "Becky" is excited by the river's response to rains, "your noisy celebration / of our celestial blessing / after years of waiting."[69] Rogelio, who has just bought his family's first house, imagines installing

> glass floors with a river—you—flowing, visible / under our feet in every room.... Reality is too present—wood, nails, tile— / repairs necessary, but a man needs to imagine / a *rio*'s hum under every room / of his first American home.[70]

It's hard to resist thinking along the lines of Heraclitus's parable declaring that one cannot step into the same river twice. Encantado's river is ever present in its various roles to the town's inhabitants—at once the river of the past that bore witness to Don Arturo's picnics and Cecilia's pregnancy, the river that assures Amy and Rogelio with its healing hum, and the river whose very existence enables the future of life in this desert place.

The penultimate poem of *Encantado* is titled "The River," and in it, readers find desert abundance. Its lines reveal the interrelationships between the human, nonhuman, and more than human, emphasized by Mora through the use of repetition:

> **In the** light is a land.
> **In the** land is a river.
> **In the** river is a song.
> **In a** city flows the song.
>
> **To the** river come the voices.
> Stories **in the** voices.
> Sorrows **in the** stories.

Longings **in the** sorrows.
Prayers **in the** longings.
Hope **in the** prayers.

Prayers **in the** hope.
Longings **in the** prayers.
Sorrows **in the** longings.
Stories **in the** sorrows.
Song **in the** stories.

Song **in the** river.
River **in the** land.
Land **in the** light.[71] (emphasis added)

The rhythmic deployment of "in the" becomes a refrain connecting light, land, voices, river, city, song, stories, prayers, hope, longings, and sorrows. Further, the repetitive prepositional phrase "in the" shifts gently from left to middle to right to middle again, visually mimicking/performing the wavy current of a river. Regarding refrains, Bennett eloquently articulates that "through the various refrains we invent, repeat, and catch from nonhumans, we receive news of the cosmic energies to which we humans are always in close, molecular proximity."[72] Mora's *Encantado*, whether via human or nonhuman speakers, pulls readers in close to her imaginary desert town. The eponymous poem titles summon the people of Encantado; their own voices tell their stories about place and identity with an intimacy not traditionally associated with desert environments. Through meeting its people, its river, its histories, its languages, and its rhythm, readers gain understanding of what the "August Bees" call its "extravagant desert sagas" and "lavish profusion of abundance."[73]

Conclusion

In both *Signs Preceding the End of the World* and *Encantado: Desert Monologues* the authors create opportunities for readers to engage desert places as dynamic and full of life, and sometimes death. These texts, as informed imaginaries, incorporate cultural practices and histories of the US-Mexico borderlands and the Desert Southwest to build speculative worlds through which desert abundance is revealed. Though they differ greatly in tone, bringing these two works together here illuminates how a range of representations of arid environments exists; Herrera's reimagining of a migrant experience through the underworlds of Mictlán conjures a violent desert hauntology, whereas Mora's diverse desert community celebrates the presence of spirits, especially on el día de los muertos, in ways that often reflect wonder and enchantment even as they channel grief. The deserts in both of these works are abundantly storied, active and alive, inviting us to interpret relationships between people and place that challenge traditional desert aesthetics of wasteland or emptiness. Unlike many canonical desert texts, *Signs* and *Encantado* don't exoticize or minimize the importance of the people in desert places. As contemporary literature that I hope I have demonstrated fits well into the relatively recent framework of Latinx environmentalisms, their representations are intimate and interpersonal reimaginings, grappling with the complexities of both human and more than human agencies—borderlands ecologies which involve politics and identities as well as the elemental and environmental—and keeping communities centered as characters contest or collaborate with one another. The storied deserts of *Signs* and *Encantado* allow us to think carefully, abundantly, about the past, present, and futures of arid environments and those who call them home.

Acknowledgments

The author would like to extend her thanks for permissions to use the poetry of Pat Mora's *Encantado: Desert Monologues* for this chapter.

Notes

1 "History & Culture – Tohono O'odham Nation," Official Web Site of the Tohono O'odham Nation, 2016, http://www.tonation-nsn.gov/history-culture/.
2 The bordering of O'odham land also now means that "O'odham bands are now broken up into federally recognized tribes: the Tohono O'odham Nation, the Gila River Indian Community, the Ak-Chin Indian Community and the Salt River (Pima Maricopa) Indian community. Each band is now politically and geographically distinct and separate. The remaining band, the Hia-C'ed O'odham, are not federally recognized, but reside throughout southwestern Arizona" ("History & Culture – Tohono O'odham Nation").
3 See Jason De León's *The Land of Open Graves: Living and Dying on the Migrant Trail* (2015), Julian Lim's *Porous Borders: Multiracial Migrations and the Law in the U.S.-Mexico Borderlands* (2017), and David Romo's *Ringside Seat to a Revolution: An Underground Cultural History of El Paso and Juárez, 1893–1923* (2005).
4 Sarah D. Wald et al., *Latinx Environmentalisms: Place, Justice, and the Decolonial* (Warszawa: Temple University Press, 2019). The editors use the term Latinx environmentalisms to "redress the vacuum of attention given to Latinx environmental thought, particularly in literature and cultural productions. [They] argue that this vacuum emerges from the colonialist and white supremacist ideologies embedded in the formulation of mainstream environmentalism" (3). I can't help but see similarities between how that vacuum was created and how the same forces have established the world's desert places through a perceived sense of void or emptiness, to the detriment of local communities.
5 Lisa Dillman translates from Spanish and Catalan. She currently teaches in the Department of Spanish and Portuguese at Emory University.
6 Yuri Herrera, *Signs Preceding the End of the World*, trans. Lisa Dillman (& Other Stories, 2015), 14.
7 This premise may additionally conjure readers' recollections of the nine circles of Hell in Dante's *Inferno*.

8 Marcelo Rioseco, "Myth, Literature, and the Border in Signs Preceding the End of the World by Yuri Herrera," trans. Arthur Malcolm Dixon, Latin American Literature Today, April 25, 2017, https://latinamericanliteraturetoday.org/2017/04/myth-literature-and-border-signs-preceding-end-world-yuri-herrera-marcelo-rioseco/.
9 Rioseco.
10 Lisa Dillman, "Translator's Note," in *Signs Preceding the End of the World*, by Yuri Herrera (& Other Stories, 2014), 111.
11 Herrera, *Signs Preceding the End of the World*, 52.
12 Herrera, 19.
13 Herrera, 65.
14 Herrera, 66.
15 Herrera, 65.
16 Dillman, "Translator's Note," 112.
17 Dillman, 112.
18 Dillman, 113.
19 Nathan Richardson, "Jarchar: Reading Makina/Makina Reading in Yuri Herrera's Señales Que Precederán El Fin Del Mundo," *Ciberletras: Journal of Literary Criticism and Culture*, no. 41 (January 2019): 19. This article provides a thorough investigation of the Spanish word and its significance. Highly recommended reading if this is of further interest.
20 Herrera, *Signs Preceding the End of the World*, 38, 50.
21 Herrera, 76.
22 Dillman, "Translator's Note," 113.
23 Ila Nicole Sheren, *Border Ecology: Art and Environmental Crisis at the Margins* (Springer Nature, 2023), 28.
24 Herrera, *Signs Preceding the End of the World*, 52.
25 This includes the consequences of a decades-long practice of Prevention Through Deterrence by the U.S. Customs and Border Patrol implemented in the 1990s. As a strategy which essentially funnels migrants into the harshest desert environments. Jason de León has written extensively about this in his book, *The Land of Open Graves*, where he writes that "nature has been conscripted by the Border Patrol to act as an enforcer while simultaneously providing this federal agency with plausible deniability regarding blame for any victims the desert may claim" (30).
26 Sheren, *Border Ecology*, 26.
27 Herrera, *Signs Preceding the End of the World*, 97, 98.
28 Herrera, 98.
29 Herrera, 99.
30 Herrera, 59–60.
31 *Oppenheimer* (Universal Pictures, 2023).
32 REG 09 US EPA, "Abandoned Mines Cleanup," Reports and Assessments, February 16, 2016, Southwest, Arizona, New Mexico, Utah, https://www.epa.gov/navajo-nation-uranium-cleanup/abandoned-mines-cleanup.
33 According to the official website, the NTS was renamed the Nevada National Security Site (NNSS) in 2010 "to more accurately reflect its evolving mission." ("About the NNSS (Nevada National Security Sites)," Nevada National Security Sites, 2023, https://nnss.gov/about-the-nnss/.)
34 Today, the NNSS carries out subcritical, or smaller underground experiments "using cutting-edge computer and predictive modeling…to ensure the nation's nuclear warheads meet performance requirements, without ever creating a sustained nuclear chain reaction." ("Smaller experiments, bigger discoveries…" 2022.)
35 John Beck, *Dirty Wars: Landscape, Power, and Waste in Western American Literature* (U of Nebraska Press, 2009), 5.
36 Herrera, *Signs Preceding the End of the World*, 11.

37 Herrera, 11.
38 Eduardo Matos Moctezuma, "'On the Threshold of Death...And Life,' from The Aztecs-Mexica and Death: A Rebirth of Gods and Men," trans. Nagao, *Artes de México*, no. 96 (2009): 68.
39 Herrera, *Signs Preceding the End of the World*, 39.
40 Herrera, 39.
41 Herrera, 43.
42 Pat Mora, *Encantado: Desert Monologues*, 1st edition (University of Arizona Press, 2018), 3.
43 Mora, 3.
44 Mora, 3.
45 Mora, 3.
46 Celina Osuna, *Desert Distortion: Revealing Desert Abundance* (Texas Tech University Press, 2025).
47 Julian Lim, *Porous Borders: Multiracial Migrations and the Law in the U.S.-Mexico Borderlands* (UNC Press Books, 2017), 2.
48 Both Emma Pérez (1999) and Vanessa Fonseca-Chávez (2020) have offered the kaleidoscope as a way to understand complexities in Chicana/o history, literature, and culture. Pérez uses it "to refer to the fragmented pieces of colonial identities produced via myriad forms of violence," while for Fonseca-Chávez it serves as a representation of the ways in which colonial legacies have functioned within Chicana/o literature. Understanding the shifts and ruptures of the kaleidoscope thus allows for a better understanding of colonial legacies that include intersectional issues of race, class, and gender" (Fonseca-Chávez 13, 8).
49 Mora, *Encantado*, 9, 15, 19, 25, 75, 73, 39, 69.
50 In addition to *Storied Deserts: Reimagining Global Arid Lands*, collections like *Deserts Are Not Empty* (2022) and *Desertscapes in the Global South and Beyond* (2023) have demonstrated increasing interdisciplinary interest in deserts around the world.
51 Cynthia Radding, *Bountiful Deserts: Sustaining Indigenous Worlds in Northern New Spain*, Latin American Landscapes (The University of Arizona Press, 2022), 10.
52 Radding defines monte as "uncultivated thorn forests" (10).
53 Gary Paul Nabhan, ed., *The Nature of Desert Nature*, Southwest Center Series (The University of Arizona Press, 2020), 14.
54 Pat Mora, *Nepantla: Essays from the Land in the Middle*, 1st ed. (University of New Mexico Press, 1993), 67.
55 Jane Bennett, *The Enchantment of Modern Life: Attachments, Crossings, and Ethics* (Princeton University Press, 2016), 5.
56 Mora, *Encantado*, 5.
57 Mora, 57.
58 Mora, 57.
59 Bennett, *The Enchantment of Modern Life*, 153.
60 Bennett, 6.
61 Mora, *Encantado*, 3.
62 Mora, 3.
63 Mora, 42.
64 Mora, 23.
65 Mora, 23.
66 Mora, 59.
67 Mora, 54.
68 Mora, 69, 75.
69 Mora, 37.
70 Mora, 73.
71 Mora, 81.
72 Bennett, *The Enchantment of Modern Life*, 168.
73 Mora, *Encantado*, 43.

Bibliography

Ach, Jada, and Gary Reger, eds. *Reading Aridity in Western American Literature. Ecocritical Theory and Practice*. Lexington Books, 2020.

Beck, John. *Dirty Wars: Landscape, Power, and Waste in Western American Literature*. U of Nebraska Press, 2009.

Bennett, Jane. *The Enchantment of Modern Life: Attachments, Crossings, and Ethics*. Princeton University Press, 2016.

De León, Jason. *The Land of Open Graves: Living and Dying on the Migrant Trail*. University of California Press, 2015.

Dillman, Lisa. "Translator's Note." In *Signs Preceding the End of the World*, by Yuri Herrera, 109–114. & Other Stories, 2014.

Fonseca-Chávez, Vanessa. *Colonial Legacies in Chicana/o Literature and Culture: Looking Through the Kaleidoscope*. Tucson, Arizona: University of Arizona Press, 2020.

Herrera, Yuri. *Signs Preceding the End of the World*. Translated by Lisa Dillman. & Other Stories, 2015.

Lim, Julian. *Porous Borders: Multiracial Migrations and the Law in the U.S.-Mexico Borderlands*. UNC Press Books, 2017.

Moctezuma, Eduardo Matos. "'On the Threshold of Death… and Life,' from the Aztecs-Mexica and Death: A Rebirth of Gods and Men." Translated by Nagao. *Artes de México*, no. 96 (2009): 66–69.

Mora, Pat. *Encantado: Desert Monologues*. 1st ed. University of Arizona Press, 2018.

———. *Nepantla: Essays from the Land in the Middle*. 1st ed. University of New Mexico Press, 1993.

Nabhan, Gary Paul, ed. *The Nature of Desert Nature. Southwest Center Series*. The University of Arizona Press, 2020.

Nevada National Security Sites. "About the NNSS (Nevada National Security Sites)." 2023. https://nnss.gov/about-the-nnss/.

Official Web Site of the Tohono O'odham Nation. "History & Culture – Tohono O'odham Nation." 2016. http://www.tonation-nsn.gov/history-culture/.

Oppenheimer. Universal Pictures, 2023.

Osuna, Celina. *Desert Distortion: Revealing Desert Abundance*. Texas Tech University Press, 2025.

Pérez, Emma. *The Decolonial Imaginary: Writing Chicanas Into History*. Theories of Representation and Difference. Bloomington: Indiana University Press, 1999.

Radding, Cynthia. *Bountiful Deserts: Sustaining Indigenous Worlds in Northern New Spain. Latin American Landscapes*. The University of Arizona Press, 2022.

Richardson, Nathan. "Jarchar: Reading Makina/Makina Reading in Yuri Herrera's Señales Que Precederán El Fin Del Mundo." *Ciberletras: Journal of Literary Criticism and Culture*, no. 41 (January 2019): 12–23.

Rioseco, Marcelo. "Myth, Literature, and the Border in Signs Preceding the End of the World by Yuri Herrera." Translated by Arthur Malcolm Dixon. Latin American Literature Today, April 25, 2017. https://latinamericanliteraturetoday.org/2017/04/myth-literature-and-border-signs-preceding-end-world-yuri-herrera-marcelo-rioseco/.

Sheren, Ila Nicole. *Border Ecology: Art and Environmental Crisis at the Margins*. Springer Nature, 2023.

US EPA, REG 09. "Abandoned Mines Cleanup." Reports and Assessments, February 16, 2016. Southwest, Arizona, New Mexico, Utah. https://www.epa.gov/navajo-nation-uranium-cleanup/abandoned-mines-cleanup.

Wald, Sarah D., David J. Vazquez, Priscilla Solis Ybarra, Sarah Jaquette Ray, and Laura Pulido. *Latinx Environmentalisms: Place, Justice, and the Decolonial*. Temple University Press, 2019.

13

"PALMAS ULTIMAS," "DESERT SURROUNDINGS," AND "SONORA, MEXICO"

Anna Flores, poetry
Elexus Lopez, photography

Palmas Ultimas

Anna Flores

In the deep gash across Kasayeo mountain's south face, the state's oldest palm trees stand together: Arizona's only indigenous frond, Washintonia Filifera.

It's said the desert's last true palm self-shears,
dropping dead fronds by way of storm. The trees
are glaciation descendants. Know
an ocean's true name. Know how desert *arrived.*
Leaf-less and blood welded, my body is
a knowing I cannot fall away from.
Words are my strained molting. I'm writing
my city's name, its tail curling. Its heart
in the ground. Language is sacred waste. It's
eaten through short roots. I'm only my mouth's
storyteller. I tell it of wildfires
two states over. Their archive, in my lung.
Another blaze today, domed into
mimic of overcast. I'm relieved at a
sick sky's shade. I'm ashamed. The sun sinks red
into twenty prophecies. Some from the
ancestors. Some from the radio
host who my mother blasts in the mornings.
He talks about the oldest, final days.

DOI: 10.4324/9781003397212-17

He prays for the sick. He recommends an
immigration lawyer. He says the end
is closer than ever before, closer
than when the oceans withered and chollas
rose from the ashes of coral reefs, closer
than death even. Closer than the Sun's truest
color. Who finds me to ask if I've seen it?
Who pays for my radioed-in blessing?
We are love to each other. We will look away
together. We'll take turns discerning between
rain clouds and burdened smoke. We'll insist on
wondering why we feel sick. We'll cough up
each other's names. We'll find the highest crevice.
We'll learn to shed what's burnt. We'll grow from there.
There, wedged in canyon washes – together.

FIGURE 13.1 “Desert Surroundings.”

FIGURE 13.2 "Desert Surroundings."

FIGURE 13.3 "Desert Surroundings."

FIGURE 13.4 "Desert Surroundings."

FIGURE 13.5 "Sonora, Mexico."

FIGURE 13.6 "Sonora, Mexico."

FIGURE 13.7 "Sonora, Mexico."

FIGURE 13.8 "Sonora, Mexico."

This visual and textual collaboration is a reflection on culture, identity, and a cyclical pattern of life set against the backdrop of a desert landscape. The poem and its accompanying photographs explore the endurance of desert Palmas and its people, impending change, and the erosion of language, ultimately highlighting the urgency of shared experiences between those like and unlike us, between rock and flesh, between the living and the dead.

CODA

Celina Osuna and Aidan Tynan

From the beginning, our approach to this book has been speculative and open. Our interest was in discovering modes of scholarship and creative practice that understand the value, diversity, and abundance of desert places and their communities. Our sense was that a worldwide interest in deserts existed among humanities and social science scholars, as well as artists, but that a recognizable field of 'desert humanities' did not, despite the rapid expansion and diversification of work in environmental humanities in recent years. As with any grand endeavor, and certainly one interested in conveying local and planetary concerns of such scale and complexity, we acknowledge there are limits to our scope and ability to represent work from each of the world's deserts and cultures. Our intention is to instigate such conversations, to nurture new currents of thinking, new approaches, and questions, to connect diverse fields and voices, and to indicate future directions for work on the Earth's arid lands in environmental humanities and beyond. We are making no claims to comprehensiveness but are proud to have brought together research and art that has such a varied range. We could never have imagined, at the outset, the breadth and quality of the work we received in response to our call for papers. We are grateful beyond words to all our contributors and regret that we could not have included more.

As we noted in our introduction, over the four years it has taken for this book to appear, a number of monographs and edited collections have been published which, together with this volume, demonstrates quite clearly that desert humanities is indeed an important emerging field. It is a field that is strikingly interdisciplinary, often spurred on by engagements between humanities, sciences, and the arts, by a sensitivity to questions of social and environmental justice, and by an awareness of how places are judged or misjudged. This is no doubt a response to the narrowness by which deserts have often been perceived—as mere backdrops to national

DOI: 10.4324/9781003397212-18

imaginaries, as signifiers of decline or empty immensity, places valued negatively against dominant conceptions of settlement and agriculture, sometimes exoticized or fetishized. One of the volume's primary goals has been to address these misapprehensions while underscoring that deserts are *peopled*, sites of lived experiences of community and tradition, where histories and futures are made. The category of the Anthropocene has had such a transformative impact on environmental humanities in recent years precisely because it demonstrates that histories of humanity and the timescales of social and cultural life can no longer be thought as separate from the slower, deeper, geological time of the Earth. Rejecting stereotypical and often damaging visions of deserts as void involves including desert places within social history rather than seeing them as forms of geographical alterity existing, somehow, outside of time. This is why we insist on *storied* deserts, stories are complex parts of interrelationships with time, space, the nonhuman, the more than human, and each other—stories are temporal, shared, told, and retold.

Elementality, poetics, and politics have been our guiding concerns. The volume has sought new ways in which the elemental materialities that have dominated desert experience—distance, rock, light, heat, sky—can be reckoned anew as components of stories of human and nonhuman belonging. At the same time, we recognize that deserts resist definition according to the facts of matter alone. They are constructed through an abundance of signs, stories, and lived experience, through the fraught historical processes by which life makes a place for itself, or is denied it. Life in deserts can, without doubt, be precarious, not simply because of the material scarcities that affect it but also, and more importantly, because their inhabitants are so often denied the rights of belonging, the rights to place. Themes of settler-colonial violence and the struggles of Indigenous peoples have figured strongly in this book and are unavoidable for desert-focused scholarship oriented around questions of social justice. Extractive capitalism and climate change have only intensified the threats to which deserts and their inhabitants have long been exposed.

Though often misrepresented or misunderstood, deserts are culturally and materially storied places. The approach of storied deserts throughout is offered up here to encourage our audience to think about the Earth's arid lands and their communities in more dynamic and fruitful ways that are not detached from the realities of the world in which we're doing research. It is our sincere hope that the explorations in this book problematize 'the desert' as a monolithic narrative, and that they invite a perspectival shift—one which generates deeper interests in desert places and cultures as vital to local and planetary futures.

INDEX

Note: Page numbers followed by "n" denote endnotes.

For Product Safety Concerns and Information please contact our EU representative GPSR@taylorandfrancis.com
Taylor & Francis Verlag GmbH, Kaufingerstraße 24, 80331 München, Germany

www.ingramcontent.com/pod-product-compliance
Ingram Content Group UK Ltd.
Pitfield, Milton Keynes, MK11 3LW, UK
UKHW020147200926
475702UK00016B/2596
* 9 7 8 1 0 3 2 5 0 1 7 9 6 *